An Introduction to Recombinant DNA in Medicine

Second Edition

Alan E. H. Emery
MD PhD DSc FRCP(E) FRS(E) FLS
Formerly Consultant Physician and
Professor of Human Genetics, University of Edinburgh, UK
Currently Research Director, European Neuromuscular Centre,
Baarn, The Netherlands

and

Sue Malcolm
MA PhD MRCPath
Reader and Head of Molecular Genetics
Institute of Child Health, University of London, UK

JOHN WILEY & SONS
Chichester · New York · Brisbane · Toronto · Singapore

Baffins Lane, Chichester,
West Sussex PO19 1UD, England

Telephone: National Chichester (01243) 779777
International (+44) 1243 779777

First edition 1984
Reprinted with corrections 1985 (twice)
Italian edition 1986
English Language Book Society edition 1987
Japanese edition 1991

Other Wiley Editorial Offices

John Wiley & Sons, Inc.. 605 Third Avenue,
New York, NY 10158-0012, USA

Jacaranda Wiley Ltd, 33 Park Road, Milton,
Queensland 4064, Australia

John Wiley & Sons (Canada) Ltd, 22 Worcester Road,
Rexdale, Ontario M9W 1L1, Canada

John Wiley & Sons (SEA) Pte Ltd, 37 Jalan Pemimpin #05-04,
Block B, Union Industrial Building, Singapore 2057

Library of Congress Cataloging-in-Publication Data

Emery, Alan E. H.
An introduction to recombinant DNA in medicine / by A. E. H. Emery
and S. Malcolm.—2nd ed.
p. cm.
Includes index.
ISBN 0 471 93984 6
1. Medical genetics. 2. Genetic engineering. 3. Recombinant DNA.
4. Pathology, Molecular. I. Malcolm, S. II. Title.
RB155.E525 1995
616'.042—dc20 95-3818
CIP

British Library Cataloguing in Publication Data

A catalogue record for this book is available from the British Library

ISBN 0 471 93984 6

Typeset in 10/12 pt Palatino by Mathematical Composition Setters Ltd, Salisbury, Wiltshire
Printed and bound in Great Britain by The Bath Press, Bath, Avon

This book is printed on acid-free paper responsibly manufactured from
sustainable forestation, for which at least two trees are planted for each one
used for paper production.

Contents

Prologue to the First Edition

After the discovery of the double-helical structure of DNA in 1953, interest in molecular biology grew considerably. The late 1950s and early 1960s were exciting times for geneticists and as a medical graduate pursuing postdoctoral studies at Johns Hopkins I was privileged to be taught by some of the best scientists in the field. Unfortunately it gradually became clear that much of what had been learned in molecular biology did not seem to bear much relevance to man, and to medicine in particular. However, interest was dramatically rekindled in the early 1970s with the advent of recombinant DNA technology or, more popularly, genetic engineering. It soon became clear that through the new technology there could be important applications in the prevention and perhaps treatment of genetic disease.

The aim of this book is to provide an introduction to the subject of recombinant DNA technology for those whose interests are perhaps more medically orientated. It is not intended to be a textbook of molecular biology, of which there are already many excellent examples, but a simple outline of the general principles of DNA technology and some of the ways in which it can be applied. In some areas the choice of examples has been highly selective, not necessarily in the belief that the data are especially illuminating, but rather because they illustrate particularly well how the technology can be applied to a certain problem. Some findings are still controversial and no doubt a few will be disproved or modified by future research. But the aim has not been to present an up-to-the-minute review of the art and the interested reader may explore the subjects discussed in greater detail by consulting the bibliography given at the end of each chapter. By emphasizing general principles, it is hoped that the reader will be helped to interpret and assess any future developments in this rapidly growing field.

In the belief that it is sometimes helpful to progress from the general to the particular, the first chapter provides an overview of the subject, put into historical perspective, and introduces many of the concepts and ideas which are dealt with in more detail in subsequent chapters. However, for those who might find this chapter at first a little daunting and too concentrated, it is suggested that it could also be read at the end and in this way provide a useful summary.

There is a sense of euphoria among molecular biologists at present, no doubt because they have sensed what Bertolt Brecht expressed so aptly:

Beauty in nature is a quality which gives the human senses a chance to be skilful.

It is hoped that the reader may also come to share some of the excitement generated by these new skills.

Preface

It is now ten years since the first edition of this little book appeared. In the intervening period the subject of molecular genetics has advanced considerably. These developments have involved almost all aspects of the field. Technologically the most important innovation has been the introduction of the polymerase chain reaction (PCR) which in many situations has now superseded Southern blotting. In regard to the application of molecular genetics, here the most impressive feature over the last few years has been the ever-increasing number of disease genes which have been located and cloned. Of the 4000 or so known disease genes, roughly a quarter have now been mapped to specific chromosomal locations. So far upwards of 200 of these have been cloned and sequenced. Most importantly, after the first edition of the book had been published, the genes for cystic fibrosis as well as Duchenne muscular dystrophy were cloned. Furthermore, the applications of this knowledge, in regard to pathogenesis and prevention through genetic counselling and prenatal diagnosis, have advanced considerably. In several instances gene therapy is now becoming a real possibility. For all these various reasons this seemed an opportune moment to revise and update the previous edition of the book. We hope it will be found interesting and useful to students of science and medicine.

Alan E. H. Emery
Sue Malcolm

Chapter 1
The development of recombinant DNA technology

Evarist Galois was a mathematical genius who was killed in an absurd duel in 1832 at the early age of 20. Yet the few notes he left behind completely transformed higher algebra. He postulated a unique theorem that could not have been understood by his contemporaries because it was based on mathematical principles which were not discovered until nearly a quarter of a century later. Similarly, and perhaps of greater relevance here, Gregor Mendel's laws of inheritance, which he first presented in 1865, were quite unique. Arthur Koestler has argued strongly in his book *The Act of Creation* (1964) that such acts of creation are very much the exception. Discoveries much more commonly involve the selection, reshuffling, combining and synthesizing of already existing facts, ideas, faculties and skills. In his memoirs (*Testimony*, 1979) a similar case has also been argued by the Russian composer Dimitri Shostakovich in the case of musical composition. According to Shostakovich every piece of music resembles other music in some way and nothing is ever totally original and unique. So when we come to consider the beginnings of genetic engineering, or more precisely recombinant DNA technology, it is difficult to pinpoint any one particular discovery which alone heralded the new science. The development of recombinant DNA technology has been very much a matter of the gradual accumulation of various discoveries, each being dependent on some previous idea or innovation. Nevertheless it is possible to single out certain events for special mention (Table 1.1).

EARLY DISCOVERIES

Perhaps the first steps towards recombinant DNA technology were taken in 1944 when Avery, MacLeod and McCarty, working at the Rockefeller Institute in New York, first showed that genetic information is stored in nucleic acid and not protein as previously believed. There are two different nucleic acids: deoxyribonucleic acid (DNA) and ribonucleic acid (RNA). DNA is found mainly in the chromosomes within the nucleus, whereas RNA occurs mainly in the nucleolus (a structure within the nucleus) as well as the cytoplasm, there being very little in the chromosomes. Genetic information is stored in DNA.

In 1953 Watson and Crick working in Cambridge proposed the double-helical structure of DNA based on their interpretation of the X-ray diffraction studies of

Table 1.1 Milestones in DNA technology

1944	Genetic information stored in nucleic acid (DNA) and not protein as previously believed
1953	Double-helical structure of DNA demonstrated
1961	First attempt to break genetic code
1966	Establishment of the complete genetic code
1970	A complete gene synthesized *in vitro*
	Discovery of the first sequence-specific restriction endonuclease, as well as the enzyme reverse transcriptase
1972	First recombinant DNA molecules generated
1973	Use of plasmid vectors for gene cloning
1975	Southern blot technique for detecting specific DNA sequences
1976	First prenatal diagnosis using a gene-specific probe
1977	Methods for rapid DNA sequencing
	Discovery of 'split genes' and somatostatin synthesized using recombinant DNA
1978	Human genomic library constructed
	Prenatal diagnosis using linkage with a restriction polymorphism
1979	Insulin synthesized using recombinant DNA
1982	Commercial production of genetically engineered human insulin
1985	DNA fingerprinting introduced
	Dystrophin gene isolated
1987	Polymerase chain reaction invented
1990	First preimplantation diagnosis carried out
1991	Expanding triplet repeats found in fragile X syndrome and subsequently other diseases
1992	Genethon publish complete human genetic map
1993	Genethon complete first human physical map based on yeast artificial chromosomes
1994	Trials of gene therapy in cystic fibrosis begun

Wilkins and Rosalind Franklin. The attraction of the proposed structure was that it accounted for the ability of the molecule to reproduce itself in such a manner that an identical replica of itself would be formed at each cell division.

DNA is composed of long chains of molecules called nucleotides. Each nucleotide is composed of a nitrogenous base (adenine, thymine, guanine and cytosine), a sugar molecule and a phosphate molecule. The arrangement of the bases in DNA is not random: guanine in one chain always pairs with cytosine in the other chain, and adenine always pairs with thymine.

At nuclear division the two strands of the DNA molecule separate and as a result of specific base pairing each chain then builds its complement (Fig. 1.1).

It seemed logical that perhaps the nitrogenous bases were in some way arranged in coded sequences, each code specifying an amino acid.

The first successful attempt to break the genetic code was made by Nirenberg and Matthaei in 1961 while working at the National Institutes of Health in the United States. Later it was shown that genetic information in the DNA molecule was stored in the form of triplet codes, i.e. a sequence of three bases determines the formation of one amino acid. By 1966 the complete genetic code for all twenty amino acids had been established. Then in 1970 Khorana and his colleagues succeeded in synthesizing a gene in the test tube for the first time by assembling its constituent base pairs.

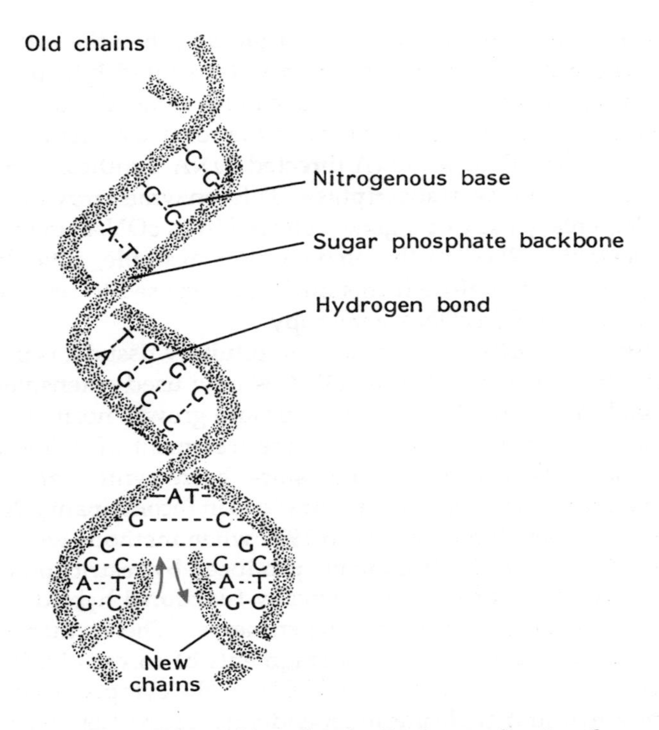

Figure 1.1 The DNA molecule. During replication the two chains separate and new complementary chains are synthesized.

There are interesting and eminently readable accounts of the discovery of the structure and function of DNA by Portugal and Cohen (1977) and Judson (1979) which are highly recommended and deal with much of the detail of this very exciting era in the history of science.

THE BEGINNING OF RECOMBINANT DNA TECHNOLOGY

Though all these early developments provided the foundation for recombinant DNA technology, the real story begins in 1970 with the discovery by Hamilton Smith of a group of enzymes which occur in microorganisms and are now referred to as class II **restriction endonucleases**. These enzymes were found to cleave the DNA molecule at sequence-specific sites, and it was this specificity exhibited by each enzyme that proved particularly important for the subsequent development of recombinant DNA technology. It meant that DNA fragments of reproducible size could be produced and that DNA fragments containing a particular gene, for example, could be cut out from the rest of the DNA molecule. Also around this time Temin, and independently Baltimore, discovered another enzyme which subsequently proved to be of immense value in recombinant DNA work. Genetic information is usually transferred from DNA to RNA (so-called messenger RNA or mRNA) by a process referred to as **transcription**, and then the mRNA migrates to the cytoplasm where its genetic information is **translated** into protein synthesis. However, Temin and Baltimore and their colleagues showed that in certain viruses genetic information could occasionally flow in the reverse direction (i.e. from RNA to DNA). This is referred to as RNA-directed DNA synthesis and the enzyme responsible is called **reverse transcriptase**. This enzyme provides a method of making a complementary copy of a gene (referred to as cDNA) in the laboratory by exposing the relevant mRNA to the enzyme. For example, reticulocytes contain only globin mRNA, and therefore if this mRNA is exposed to reverse transcriptase it is possible to make a globin cDNA gene copy.

Conversely, by using mRNA extracted from pituitary tissue it is possible to make the gene for growth hormone. Globin cDNA is now used extensively to diagnose various haemoglobin abnormalities in the fetus and growth hormone cDNA is used in the biosynthesis of growth hormone for the treatment of children with certain types of dwarfism. Many cDNAs have since been synthesized using reverse transcriptase, and these find very wide applications in recombinant DNA technology.

The next important development was in 1972 when methods were first described for joining (recombining) DNA fragments, produced by restriction enzymes, from two different organisms to produce biologically functional hybrid DNA molecules – the first successful recombinant experiments. There then followed the introduction of **plasmid vectors** to carry fragments of foreign DNA. Plasmids are naturally occurring small circular pieces of DNA in the cytoplasm of bacteria which are extrachromosomal and replicate independently of the host bacterial DNA. By using a particular endonuclease the plasmid can be opened up and a foreign piece of DNA, digested with the same enzyme, inserted. This is in fact the actual **recombinant** part of the process. In 1973 Stanley Cohen constructed the first plasmid used in recombinant DNA work and which is referred to as pSC101 (p for plasmid, SC after its originator) and later other plasmids were developed specially tailored to make growing, selecting and sequencing easy.

These plasmids provide useful vehicles for carrying DNA fragments or genes, and when taken up by a bacterium (usually *Escherichia coli*) the latter can then be grown in culture to produce clones with multiple copies of the incorporated DNA fragment. Picking a single bacterial colony is the actual **cloning** part of the process.

In 1975 E.M. Southern, working in Edinburgh, introduced another important technique which proved invaluable for subsequent developments in the field. If DNA is extracted from a tissue, say leucocytes or amniotic fluid cells, this DNA contains all the thousands of genes of the host organism, though of course in any one tissue most of these are switched off or repressed and only a relative few are active. How then can a particular gene or nucleotide sequence be identified among all this DNA? The first step is to cleave the DNA with a restriction endonuclease. The resultant fragments can then be subjected to electrophoresis on agarose gel, a technique which results in the fragments separating under an electric current according to their size – smaller fragments migrating further than larger fragments. Among all these fragments will be the one containing the gene of interest. Southern's technique, details of which will be dealt with later, involved several steps. Firstly, the DNA fragments on the gel are denatured with alkali. The denaturing process renders the normally double-stranded fragments single-stranded so that they will subsequently hybridize with similar complementary pieces of DNA. In molecular biology the term **hybridize** means the sticking together to form a double-stranded molecule, either of two single-stranded DNA molecules or of a DNA and an RNA molecule. Secondly, the denatured DNA fragments are transferred to a nylon filter by in effect blotting – the filter being laid over the gel. The DNA fragments then become firmly bound to the filter. This is now universally referred to as a **Southern blot** and the beauty of the technique is that it makes it possible to fix DNA fragments on the filter at the exact positions they occupied on the original gel – this position being determined by their size. Because the fragments are denatured they will hybridize to a probe (page 46) containing DNA complementary to the particular DNA fragment of interest. When the probe is made radioactive then it is possible to identify and localize the fragment on the filter by **autoradiography** – i.e. by exposing the filter to an X-ray film in the dark.

PRENATAL DIAGNOSIS

Over the last few years the practice of medical genetics has been revolutionized by the advent of carrier detection and prenatal diagnosis. When one of the parents is shown to be a carrier of a genetic disorder they may be offered prenatal diagnosis, whereby many genetic disorders in the fetus can be detected in early pregnancy. If the fetus proves to be affected the parents may opt to have the pregnancy terminated. On the other hand, if the fetus proves to be unaffected the pregnancy is allowed to continue to term. Many couples are saved from unnecessary worry and intervention by establishing that the parent at risk is not in fact a carrier and so there is no risk to the child. Suitable samples for prenatal diagnosis may be obtained from amniotic fluid or by chorionic villus biopsy. Cells may be cultured to allow chromosome abnormalities or inborn errors of metabolism to be detected. The advent of DNA technology opened up an entirely new possibility – of detecting an abnormal gene in the DNA from the cells. Chorionic villus biopsy has the advantage that a sample may be safely taken at 10 weeks of pregnancy, allowing diagnosis to be completed before the pregnancy is generally obvious.

Using the DNA technology described so far, the first prenatal diagnosis of a genetic disease was made in 1976 by Kan, Golbus and Dozy. They were able to show that if the fetus was affected by α-thalassaemia, a disorder in which globin genes are absent (deleted), then there was reduced hybridization with an appropriate globin cDNA probe prepared from reticulocyte mRNA using reverse transcriptase.

This represented a monumental step forward in the field of prenatal diagnosis. In the case of haemoglobinopathies it was possible to replace the dangerous (to the fetus) technique of fetal blood sampling with early chorion biopsy. It paved the way for the later developments in which mutations in many genes including those causing Duchenne and Becker muscular dystrophy, cystic fibrosis and myotonic dystrophy could be detected in early pregnancy. These genes were isolated in 1985, 1989 and 1991 respectively.

In 1990 the first prenatal diagnosis was carried out on an embryo prior to implantation when twin girls, at risk for the X-linked disorder Duchenne muscular dystrophy, were born following preimplantation sexing.

GENE STRUCTURE

In 1977, a year after Kan's report, there appeared two new, and relatively simple, methods for rapidly sequencing the bases in any DNA molecule. One of the methods was developed at the MRC Laboratory of Molecular Biology in Cambridge by Sanger, Nicklen and Coulson and the other at Harvard University by Maxam and Gilbert. The DNA sequencing methods evolved by these workers meant that given a piece of DNA it was now possible to analyse its detailed structure and this was to have far-reaching consequences.

It was also at this time that recombinant DNA work led to a major discovery concerning gene structure. It had always been assumed that genes were discrete and contiguous stretches of DNA which coded for particular enzymes or peptides. However, in 1977, and here the first report in higher organisms concerned the globin gene of the rabbit, it was found that genes were in fact rarely continuous stretches of DNA. It turned out that at least in higher organisms almost all genes are interrupted by so-called **intervening sequences** or **introns**, the remaining parts of the gene separated by introns being called **exons**. This led to the concept of **split** genes. Subsequently, it was shown that during transcription the precursor RNA derived from introns is excised (for reasons which are still not clear) and the precursor RNA from non-contiguous exons is spliced together to form functional mRNA. Thus exons, and not introns, specify the primary structure of the gene product. This finding was also to have far-reaching consequences, particularly with regard to the use of DNA technology in biosynthesis.

BIOSYNTHESIS

At the same time as the fine details of gene structure were beginning to be revealed, the first biosynthesis of a human protein using DNA technology was reported by Itakura, and incidentally the first genetic engineering company (Genentech Inc. in the United States) was founded specifically to develop recombinant DNA methods for making medically important drugs. Itakura and his colleagues successfully obtained the bacterial expression of a cloned gene for somatostatin, a peptide hormone which, among other things, inhibits growth hormone and is used in the treatment of children with excessive growth. From knowing the amino acid structure of somatostatin they were able to infer the nucleotide base composition of its gene. This was relatively easy because somatostatin contains only 14 amino acids. The assembled synthetic gene, along with various gene regulators, was then inserted into the plasmid pBR322 and cloned in the bacterium *Escherichia coli*. Since that first step many other valuable hormones and proteins have been synthesized using DNA technology. For example, Goeddel and his colleagues at Genentech published details of their successful synthesis of human insulin in 1979, though in fact the expression of human insulin in *E. coli* was announced on 6 September 1978 – an auspicious day. Subsequently, marketing approval for genetically engineered insulin was obtained by the Eli Lilly Company in 1982 and is now freely available for the treatment of diabetes.

GENE LIBRARIES

As more experience was gained in cloning DNA sequences and confidence in the techniques grew, the feasibility of possibly cloning DNA sequences from the entire human genome began to be entertained. DNA fragments produced by restriction enzymes and cloned in appropriate vectors could be stored and in this way a so-called **library** of an individual's genome could be produced. The pioneers in this regard were Maniatis and his colleagues at Cal. Tech. who reported the first successful construction of a human gene library in 1978. This was to prove extremely valuable for subsequent research.

RESTRICTION POLYMORPHISMS

In the same year, 1978, there was also published a paper which heralded yet another major advance, again in the field of prenatal diagnosis. Recombinant DNA technology had so far been used to detect genetic disorders in the fetus by demonstrating an abnormality in amniotic fluid cell DNA using a gene-specific probe. However, in many genetic disorders, the basic biochemical defect is unknown and the production of gene-specific probes is very difficult or indeed not possible. Kan and Dozy in 1978 showed that an entirely different approach was feasible. The new approach depended on demonstrating that the disease-producing gene is very close (closely linked) to a particular restriction enzyme recognition site. It is known that variations in DNA sequences occur randomly throughout the entire genome and are apparently without any apparent ill effects. These variations (base changes) result in the loss of an existing restriction site or the acquisition of a new site. Such changes in DNA sequences mean that the fragments produced by a particular restriction enzyme will be of different lengths in different people, and can be recognized by their different mobilities on electrophoresis. They are referred to as **restriction fragment length polymorphisms** (RFLPs) and are inherited as simple genetic traits obeying the Mendelian laws of inheritance. If a disease-producing gene could be shown from family studies to be closely linked to an RFLP then it could provide a means of detecting the disease-producing gene without actually knowing anything at all about the gene itself.

Kan and Dozy found that with a particular restriction enzyme (*Hpa*I), DNA fragments incorporating the β-globin gene were of different sizes depending on whether the chromosome carried a normal globin gene or the gene for sickle cell anaemia. In their experiments these investigators used a gene-specific probe and the DNA fragments produced contained the gene of interest, i.e. the β-globin gene. But neither of these is necessary in this approach. In principle all that is necessary is to find close linkage between a particular RFLP and a disease-producing gene.

ONCOGENES

1982 saw the publication of the first attempts to isolate, clone and characterize a human cancer gene – in this case a gene associated with bladder cancer. Essentially the technique used by these and other investigators is to extract DNA from cancer tissues, fragment the DNA with a restriction enzyme and then apply these fragments to a special line of cultured mouse fibroblasts, referred to as NIH 3T3. In tissue culture these cells become transformed and form clumps (rather than a flat single layer of cells) when exposed to whole tumour cell DNA from cancer tissues. What these and other investigators have found is that certain **fragments** of DNA also have the same effect, and that DNA fragments capable of transforming NIH 3T3 cells contain cancer genes or **oncogenes**. We all carry proto- or non-mutated forms of these genes. To paraphrase Molière, in the *Bourgeois Gentilhomme*, we have been inheriting oncogenes all our lives without knowing. One of the principal ways in which they become activated is by chromosome translocation. Subsequently the genes involved in a number of familial cancers have been isolated. An important class turn out to be tumour suppressor genes, which normally hold the growth of the cell in check, but lose control when mutated. In 1993 it was discovered that the main gene causing colon cancer is a mutator gene, normally keeping errors of DNA replication under surveillance and in 1994 a gene predisposing to breast cancer, carried by around 1 in 200 persons, was identified.

THE POLYMERASE CHAIN REACTION

The technology for handling DNA was revolutionized in 1985 with the invention by Kary Mullis of the polymerase chain reaction (PCR) or thermal cycling. This used simple enzyme reactions and amplified specifically, from the 3 billion base pairs of DNA in each cell, a stretch of a few hundred base pairs, until there were hundreds of thousands of identical copies of the starting sequence. To start the reaction two oligonucleotides (or primers) of around 20 base pairs are synthesized corresponding to DNA sequences on either side of the stretch to be amplified. The starting DNA is heated to separate the strands and then cooled in the presence of the oligonucleotides. By base pairing they will form double-stranded molecules on either side of the target. DNA polymerases, in the presence of nucleotide triphosphates, will use these short double-stranded sections as templates and will fill in the second strand, which will of course be an exact copy of the original. After one cycle there will be two copies of the target region. After further cycles of heating, annealing and copying there will be an exponential increase in the number of copies of the target. Theoretically, and in practice if great care is taken, the PCR reaction can amplify a DNA sequence from a single cell. It has made DNA analysis possible on minute samples of poor quality DNA including human remains, forensic samples, paraffin blocks and even reached popular culture with descriptions of recreating dinosaurs.

MICROSATELLITES AND LENGTH POLYMORPHISMS

As the analysis of human DNA has progressed it has become obvious just how variable the sequences are between individuals. Microsatellites were first detected by Alec Jeffreys in 1985. These are runs of short sequences which are tandemly repeated. Many such satellites exist and they are scattered over all human chromosomes. Their repetitive nature makes them unstable during DNA replication between generations and slippage or stuttering sometimes occurs. The consequence is that the number of repeats within the run varies widely between individuals. If variation at several sites is taken into consideration each individual, except identical twins, will have a personal DNA fingerprint or profile. These profiles have been used to establish individual identity in paternity, immigration, forensic and clinical cases.

In 1989 Mike Litt and Jeffrey Luty, and Jim Weber and Paula May published simultaneously their observation that the commonest satellite sequence in human DNA consists of a run of the dinucleotides CA or GT, depending on which strand of the DNA is being read, and that there are many thousands of these scattered apparently randomly throughout human DNA. The length of the repeat can be highly variable and they provide excellent markers for tracking through families. They have largely replaced the RFLPs described above and a mammoth effort, using industrial processes, enabled the workers at Genethon outside Paris to publish the first complete genetic map of the human genome in 1992.

Most of these microsatellite sequences produce no harmful effect but in 1991 it was shown that a form of mental retardation in males, the fragile X syndrome, was caused by the expansion of a triplet repeat (CGG) from 6 to 54 repeats in the normal range to over 200 in affected boys. Other late onset neurological disorders, in particular myotonic dystrophy and Huntington's disease, were later found to arise by a similar mechanism. This discovery provided a scientific basis for the phenomenon of anticipation in which some genetic disorders were observed to become more severe in succeeding generations.

SUMMARY

This brief historical introduction has provided an opportunity to consider some of the general and more important aspects of recombinant DNA technology.

- Gene specific probes may be synthesized chemically, made from cDNA, made from fragmented genomic DNA using a restriction endonuclease or amplified using sequence specific primers in the polymerase chain reaction.
- Such probes can be used to detect harmful genes in preclinical cases and healthy carriers of serious genetic disorders as well as in prenatal diagnosis.
- Cloned genes may also be used in the biosynthesis of medically important substances such as insulin and the earliest steps are being taken in gene therapy.
- The human genome is now densely covered with genetic markers so that almost any disease can be mapped. This is often a first step towards isolating the gene.
- We have the technology to study the detailed fine structure of genes themselves and to detect mutations in individual patients (Fig. 1.2). Most interestingly, we can determine the basis of genetic disease by studying mutant genes directly and discover previously unknown protein products.

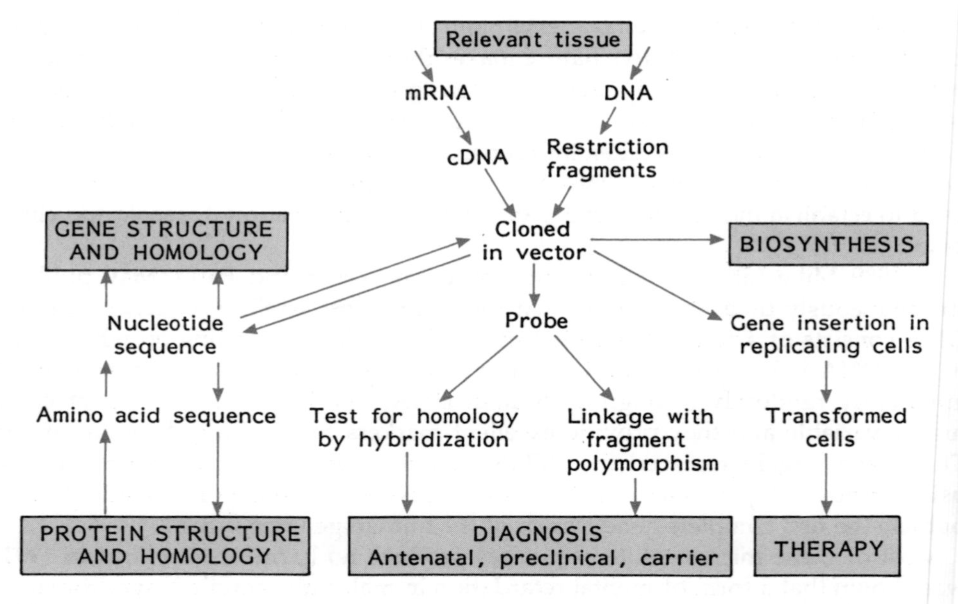

Figure 1.2 Some of the applications of DNA technology and how these are approached.

REFERENCES AND FURTHER READING

Textbooks and review articles

Judson HF. *The Eighth Day of Creation – Makers of the Revolution in Biology*. New York: Simon & Schuster, 1979

Portugal FH, Cohen JS. *A Century of DNA: A History of the Discovery of the Structure and Function of the Genetic Substance*. Cambridge, Mass.: MIT, 1977

Shapiro R. *The Human Blueprint*. London: Cassell, 1992

Research publications

Baltimore D. RNA-dependent DNA polymerase in virions of RNA tumour viruses. *Nature* 1970; **226**: 1209–1211

Bolivar F, Rodriguez RL, Greene PJ, Betlach MC, Heyneker HL, Boyer HW. Construction and characterization of new cloning vehicles. II. A multipurpose cloning system. *Gene* 1977; **2**: 95–113

Cohen SN, Chang ACY, Boyer HW, Helling RB. Construction of biologically functional bacterial plasmids *in vitro*. *Proc Natl Acad Sci USA* 1973; **70**: 3240–3244

Duesberg PH. Retroviral transforming genes in normal cells? *Nature* 1983; **304**: 219–226

Goeddel DV, Kleid DG, Bolivar F, Heyneker HL, Yansura DG, *et al*. Expression in *E. coli* of chemically synthesized genes for human insulin. *Proc Natl Acad Sci USA* 1979; **76**: 106–110

Handyside A, Kontogianni EH, Hardy IC, *et al*. Pregnancies from biopsied human pre-implantation embryos sexed by Y-specific DNA amplification. *Nature* 1990; **344**: 768–770

Itakura K, Hirose T, Crea R, Riggs AD, Heyneker HL, *et al.* Expression in *E. coli* of a chemically synthesized gene for the hormone somatostatin. *Science* 1977; **198**: 1056–1063

Jackson DA, Symons RH, Berg P. Biochemical method for inserting new genetic information into DNA of Simian virus 40: circular SV40 DNA molecules containing lambda phage genes and the galactose operon of *E. coli. Proc Natl Acad Sci USA* 1972; **69**: 2904–2909

Jeffreys AJ, Flavell RA. The rabbit β-globin gene contains a large insert in the coding sequence. *Cell* 1977; **112**: 1097–1108

Jeffreys AJ, Wilson V, Thein SL. Hypervariable 'minisatellite' regions in human DNA. *Nature* 1985; **314**: 67–73

Kan YW, Dozy AM. Antenatal diagnosis of sickle-cell anaemia by DNA analysis of amniotic fluid cells. *Lancet* 1978; **ii**: 910–912

Kan YW, Golbus MS, Dozy AM. Prenatal diagnosis of α-thalassemia. Clinical application of molecular hybridizations. *N Engl J Med* 1976; **295**: 1165–1167

Koenig M, Monaco AP, Kunkel LM. The complete sequence of dystrophin predicts rod-shaped cytoskeletal protein. *Cell* 1988; **52**: 219–228

Koenig M, Hoffman EP, Berteloon CJ, Monaco AP, Kenner C, Kunkel LM. Complete cloning of the Duchenne muscular dystrophy (DMD) gene and preliminary genomic organisation of the DMD gene in normal and affected individuals. *Cell* 1987; **50**: 509–517

Lawn RM, Fritsch EF, Parker RC, Blake G, Maniatis T. The isolation and characterization of linked δ and β-globin genes from a cloned library of human DNA. *Cell* 1978; **15**: 1157–1174

Litt M, Luty JA. A hypervariable microsatellite revealed by in vitro amplification of a dinucleotide repeat within the cardiac muscle actin gene. *Am J Hum Genet* 1989; **44**: 397–401

Maxam AM, Gilbert W. A new method for sequencing DNA. *Proc Natl Acad Sci USA* 1977; **74**: 560–564

Reddy EP, Reynolds RK, Santos E, Barbacid M. A point mutation is responsible for the acquisition of transforming properties by the T24 human bladder carcinoma oncogene. *Nature* 1983; **300**: 149–152

Saiki R, Scharf S, Faloona F, *et al.* Enzymatic amplification of β-globin genomic sequences and restriction site analysis for diagnosis of sickle cell anaemia. *Science* 1985; **230**: 1350–1354

Sanger F, Nicklen S, Coulson AR. DNA sequencing with chain-terminating inhibitors. *Proc Natl Acad Sci USA* 1977; **74**: 5463–5467

Southern EM. Detection of specific sequences among DNA fragments separated by gel electrophoresis. *J Mol Biol* 1975; **98**: 503–517

Vererk AJMH, Pieretti M, Sutcliffe JS, *et al.* Identification of a gene (FMR-1) containing a CGG repeat coincident with a break-point cluster region exhibiting length variation in fragile X syndrome. *Cell* 1991; **65**: 905–914

Watson JD, Crick FHC. Molecular structure of nucleic acids – a structure for deoxyribose nucleic acid. *Nature* 1953; **171**: 737–738

Weber JL, May PE. Abundant class of human DNA polymorphisms which can be typed using the polymerase chain reaction. *Am J Hum Genet* 1989; **44**: 388–396

Weissenbach, Gyapy G, Dib C, *et al.* A second generation linkage map of the human genome. *Nature* 1992; **359**: 794–801

Chapter 2
Structure and function of DNA

Before discussing recombinant DNA technology in any detail it is first necessary to consider the structure of DNA itself as well as the processes of transcription and translation of genetic information. Detailed treatments of these subjects can be found in several excellent texts listed at the end of the chapter. Here we shall only be concerned with those general principles which relate particularly to recombinant DNA technology.

STRUCTURE OF DNA

Nucleic acids consist of chains of **nucleotides**. Each nucleotide contains a nitrogenous base, a five-carbon (pentose) sugar and a phosphate group. There are two types of pentose sugars found in nucleic acids: 2-deoxyribose in DNA (deoxyribonucleic acid) and ribose in RNA (ribonucleic acid). The difference is in the absence or presence of a hydroxyl group at position 2 of the sugar molecule (Fig. 2.1).

The nitrogenous bases are also of two kinds: **pyrimidines** with a six-member ring and **purines** with fused five- and six-member rings. The pyrimidines are cytosine, uracil and thymine, and the purines are adenine and guanine. For convenience they are often referred to by their initial letters (C, U, T, A and G). DNA contains adenine, thymine, guanine and cytosine. RNA also has adenine, guanine and cytosine but thymine is replaced by uracil. The only difference between thymine and uracil is the absence of a methyl group in uracil. The base-sugar moiety is called a **nucleoside** and the base–sugar–phosphate moiety is called a **nucleotide**. The nitrogenous bases and their equivalent nucleosides and the abbreviations used for their nucleotides are given in Table 2.1.

The nucleotides are linked together to form polynucleotide chains. In general RNA consists of a single polynucleotide chain, whereas DNA (unless denatured) is composed of two polynucleotide chains arranged in a double helix. Otherwise the primary structures of DNA and RNA are the same. By convention the numbers of the carbon atoms in the pentose sugars in nucleic acids are given a prime ('). In the sugar–phosphate backbone of each polynucleotide chain, the 5' position of one pentose sugar is connected to the 3' position of the next pentose sugar via a phosphate group. The terminal nucleotide at one end of the chain has a free 5' group while the terminal nucleotide at the other end has a free 3' group. The two polynucleotide chains in DNA run in opposite directions and are said to be **antiparallel**.

Projecting inwards from the two sugar–phosphate backbones are the nitrogenous bases which exhibit specific base pairing: guanine in one chain always pairs with cytosine in the other chain and adenine always pairs with thymine. The two chains are held together by hydrogen bonds between the nitrogenous bases. The important structural features of the DNA molecule are shown diagrammatically in Fig. 2.1.

2-deoxyribose *Ribose*

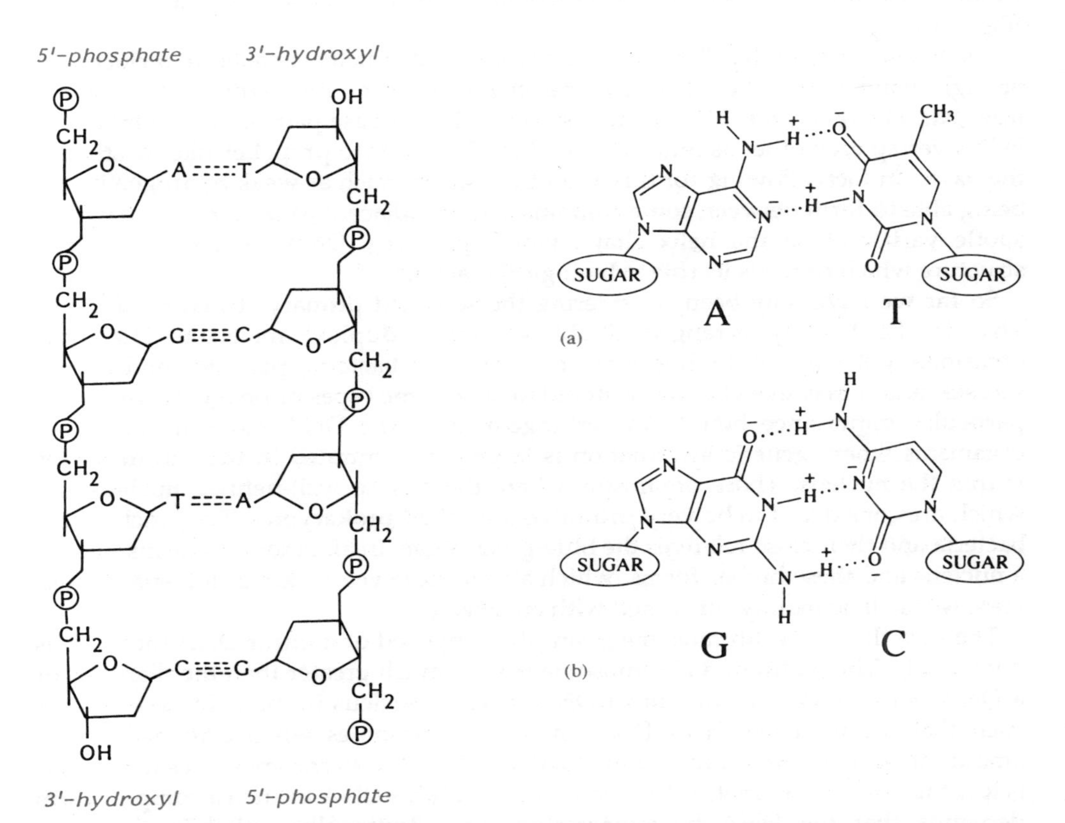

Figure 2.1 (*left*) Diagrammatic representation of the antiparallel arrangement of the two polynucleotide chains in DNA. P represents a phosphate group. (*right*) DNA bases showing hydrogen bonding in Watson–Crick base pairs: (a) A–T; (b) G–C. Symbols +, – are partial changes.

Table 2.1 Nitrogenous bases and their equivalent nucleosides and the abbreviations used for their nucleotides, e.g. AMP = adenosine monophosphate etc. ('d' represents the deoxy form in DNA)

		Abbreviation of nucleotide	
Base	Nucleoside	DNA	RNA
Adenine	Adenosine	dAMP	AMP
Guanine	Guanosine	dGMP	GMP
Cytosine	Cytidine	dCMP	CMP
Thymine	Thymidine	dTMP	–
Uracil	Uridine	–	UMP

Because of the specific base pairing, during cell division each strand acts as a template for the synthesis of a complementary strand, as we have already seen (Fig. 1.1).

The major, or so-called **B-form** of the DNA double helix is coiled in a clockwise or right-handed direction because the chains turn to the right as they move upwards. However, the DNA is more stable if the flat base pairs in the centre of the helix overlap each other as much as possible. This leads to propeller-like twisting of the bases. In fact, allowing for this and other factors such as weak hydrogen bonds being able to form between some combinations of adjacent base pairs, there will be subtle variations in the helix shape which give any DNA sequence a unique structure which controls its role in biological reactions.

So far we have only been considering the so-called primary structure of DNA. That is the linearly arranged double-stranded (duplex) molecule. However, occasionally it may not be linear but circular as in bacteria, plasmids and certain viruses, and it may even be single-stranded as in some types of phage. However, of particular importance here is the arrangement of the DNA molecule in higher organisms where genetic information is largely concentrated in the chromosomes within the nucleus. Those organisms where there is no well-defined nucleus and which are considered to be very primitive are called **prokaryotes**. They include the bacteria and their close relatives the blue-green algae. In all other organisms there is a nucleus and these higher forms, which also include yeasts, are called **eukaryotes**. Here we shall be mainly concerned with eukaryotes.

The idea that each chromosome is simply composed of a linear DNA molecule is not tenable. The width of a chromosome is very much greater than the diameter of a DNA helix. Also, the amount of DNA in each nucleus in man, for example, is such that the total length of DNA in his chromosomes when extended would amount to several metres. In fact the total length of the 46 chromosomes in humans is less than half a millimetre. To accommodate all this DNA in such a compact form demands that the DNA be **supercoiled**, i.e. additionally coiled in the same clockwise direction as the instrinsic winding of the DNA helix.

Evidence suggests that there are probably four levels of coiling: primary coiling of the DNA duplex itself; secondary coiling around the outside of spherical histone

(DNA-binding protein) 'beads' forming what are called nucleosomes; tertiary coiling with the nucleosomes forming a cylinder or solenoid-like structure; and finally quarternary coiling in the form of loops (Fig. 2.2). This supercoiled arrangement is the natural form of DNA. Energy is therefore required to unwind the DNA molecule and particular enzyme systems can accomplish this during cell division and when sections of the DNA are to be transcribed. One of the enzymes involved in the uncoiling of DNA has been given the intriguing name of swivelase,

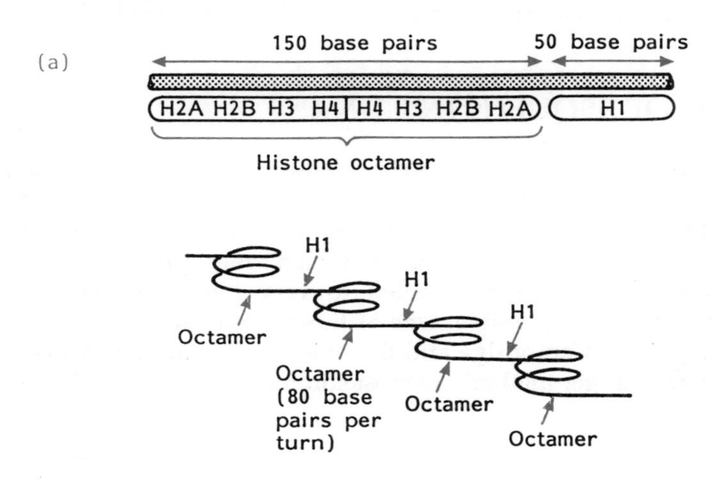

Figure 2.2 (a) DNA winds into nucleosomes around histone octamers. (b) Chain winds into coil. (c) Long linear molecule is divided into loops by attachment to scaffold (based on Callardine and Drew, 1992).

though this has now been replaced by the less evocative but more scientific term DNA topoisomerase. Because of the deleterious effect on DNA replication attempts have been made to use topoisomerase inhibitors as anti-cancer drugs. A great deal is already known of the processes involved in the coiling and uncoiling of DNA in prokaryotes but in eukaryotes much is still not clear. During programmed cell death, or apoptosis, DNA in the nucleus is broken down very quickly. The degrading enzymes cut between the nucleosomes and the resulting 'ladder' of DNA pieces is a sign that programmed cell death has occurred.

METHODS OF SEQUENCING DNA

It is often necessary to know the precise sequence of nucleotide bases in a DNA molecule. A number of techniques for doing this have been described but unfortunately the early methods were slow and laborious. However, in 1977 two new methods were introduced, both of which were quicker and more accurate. What is more, they could be applied to very long stretches of DNA which previously had not been possible. One of these methods was developed by Sanger and his colleagues at Cambridge and the other by Maxam and Gilbert at Harvard. Variations on the Sanger method are most commonly used nowadays.

Sanger method

Essentially this method involves using one strand of a DNA molecule as a template to synthesize a complementary strand, but in the synthesis the normal nucleotides (deoxynucleoside triphosphates) are replaced by their dideoxy analogues. The latter are the same as the normal nucleotides except they lack 3' hydroxyl groups. As a result, when they become incorporated into a growing DNA chain they act as terminators because the end of the chain no longer has a free 3' hydroxyl group and so no other nucleotide can be added. To aid the synthetic process a primer is added. This is a short sequence of bases which is complementary to part of the template. It may be either synthesized chemically or produced by a restriction enzyme.

The template and primer are first denatured, rendering them single-stranded. They are then annealed together to form a template–primer complex which is divided into four samples, to each of which is added DNA polymerase. One sample is incubated with a mixture of ddGTP (dideoxyguanosine triphosphate) and dGTP, together with the other three deoxynucleoside triphosphates, one of which is labelled with radioactive phosphorus (^{32}P). As the DNA chains are added to the 3' end of the primer the sites of guanine (G) are filled in by the normal dG and extended further, but occasionally by ddG and terminated. Similar incubations are carried out in the presence of each of the three other dideoxynucleotides, resulting in chains which terminate at C, A or T. The four mixtures are then subjected to gel electrophoresis, the smaller fragments migrating further than the larger fragments. Since all the chains in each of the four mixtures will end with one of the four appropriate chain terminators, the sequence of bases can simply be read off from an autoradiograph of the gel (Fig. 2.3).

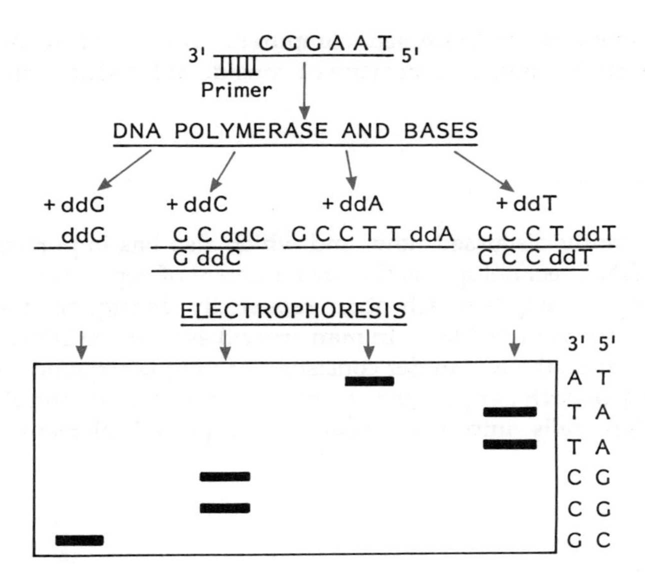

Figure 2.3 DNA sequencing method of Sanger using chain-terminating inhibitors.

A drawback to Sanger's method is that it requires single-stranded DNA as a template. One useful way round this is to label the template at one end with biotin covalently bound to a nucleotide. This strand can be separated by mixing with magnetic beads coated with avidin (avidin and biotin form a very strong complex). All the wash procedures can be carried out using a magnet to separate the DNA coated beads.

Increasingly the process is becoming automated. Each nucleotide can be labelled with a different fluorescent dye, for example FAM which is blue, JOE green, TMR yellow and ROX red, and all four samples loaded into the same gel lane. A laser is used to record the order in which the coloured molecules pass down a gel.

These methods have been extensively used in recent years. So straightforward are the methods that in many laboratories it has become an accepted practice for graduate students to sequence a particular DNA molecule as part of their project. Coupled with computer analysis of the banding patterns obtained by these methods it has been possible to completely sequence the DNA in specific genes, and to compare the sequence in normal genes, and those of patients.

It is one of the aims of the Human Genome Project (Collins and Galas, 1993) that entire human DNA is sequenced, in which the haploid genome consists of 3 thousand million (3×10^9) base pairs or 3×10^6 kilobases (kb) where 1 kb = 1000 base pairs. 2005 is the target date for this. Other targets of the Genome Project have been achieved ahead of schedule.

Knowing the sequence of bases alone, however, tells us little. What is more important is the identification of those sequences concerned with gene function. The recognition of those sequences which are not interrupted by stop codons and code for proteins (so-called **reading frames**) among the many hundreds or even thousands of bases in any given stretch of DNA is being made easier by the application of specially programmed computers. The wedding of recombinant DNA and computer technology is proving of considerable value in this field.

REPETITIVE DNA

A matter conveniently discussed now, and which also has important implications in recombinant DNA technology, is the phenomenon of repetitive DNA. These are sequences of varying length which occur repeatedly throughout the genome. All eukaryotes possess repetitive DNA. In man around 40% of the DNA is represented as repeated sequences, the remainder consisting of unique sequences where there is only a single copy of each per genome and this includes the structural genes.

Repetitive DNA falls into two classes: interspersed elements and tandem repeats.

Interspersed elements

A few gene families, including the histocompatibility genes and immunoglobulin genes, consist of closely related sequences. Many more copies, around 100 000 or more, exist of families of repeats first revealed by, and therefore named after, certain restriction enzymes. The best characterized is the *Alu* family. The *Alu* sequences are about 300 base pairs long and have obviously arisen by duplication of an original sequence of half the length as the sequence is repeated in the right and left halves of the molecule (Fig. 2.4). At one end there is a run of As and at each

(a)

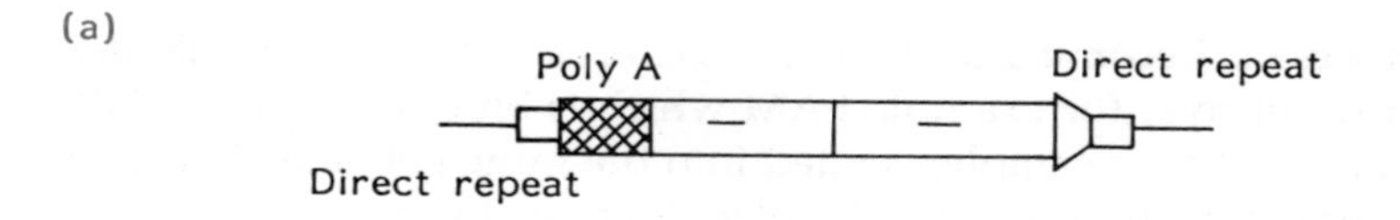

(b)

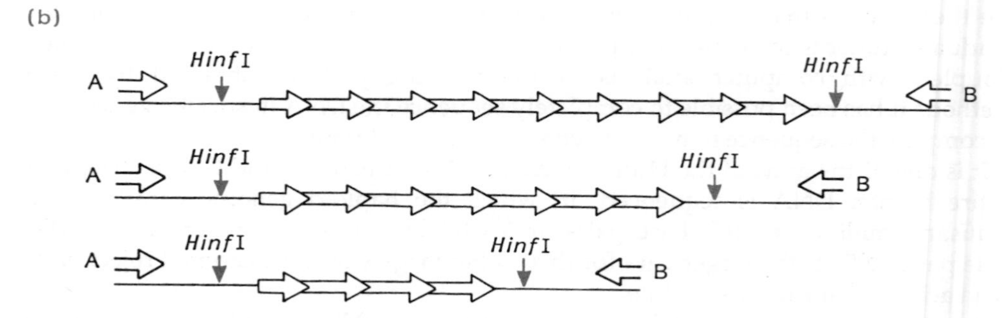

Figure 2.4 (a) Structure of interspersed *Alu* repeat. (b) Structure of tandem repeats. Three different alleles are shown with 9, 7 or 4 repeat units (arrows). The different size fragments can either be visualized using a restriction enzyme (here *Hinf*I) to cut in the surrounding sequence and detection of the fragments by Southern blotting, or by PCR amplification using flanking primers (here A and B).

end of the sequence there is a short repeated sequence of DNA. Together with the random distribution of *Alu* sequences throughout the genome the structure indicates that the the *Alu* sequences are transposable elements which have spread through the genome via an RNA intermediate (because of the stretch of As which are added to RNA after transcription). During recent studies of the gene which causes Huntington's disease an *Alu* repeat was found in two patients. This shows that some *Alu* sequences are still active in transposition. Rodents do not contain *Alu* repeats but have their own Short Interspersed Nuclear Elements (SINEs), B1, which is only half of the *Alu* dimer. This suggests that the *Alu* repeat duplicated after the separation during evolution of mice and men and the copies have spread throughout onto human DNA since then.

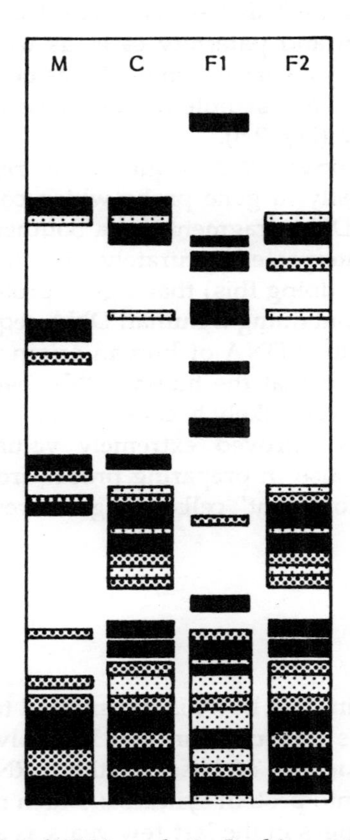

Figure 2.5 A real life example of DNA analysis. In this particular case, it was not known which of two possible fathers was the real father of the child and DNA fingerprint tests were requested to resolve the matter. Two complementary multi-locus DNA probes were used. A section of the results from one of the two probes is illustrated. The DNA fingerprint pattern of the second man (F2) contains all 20 of the paternal bands identified in the child (C), whereas that of the first man (F1) only contains 4 such paternal bands. F2 is therefore the true biological father of that child. The probability of 20 such bands matching by chance in anyone but the true father is some 1 million million to 1 against. M = pattern from mother; C = pattern from child; F1, F2 = pattern from possible fathers. (Reproduced by permission of Cell Mark Diagnostics, Blacklands Way, Abingdon, Oxfordshire, UK.)

Tandem repeats

Tandem repeats (see Fig. 2.4) are widespread in human DNA. As with interspersed repeats they vary widely in their number and complexity. The centromeres of human chromosomes have alphoid repeats up to 5 million base pairs (5 Mb) long. Amongst the simplest repeats are stretches of di, tri or tetra nucleotides. The number of units in the repeat generally varies causing changes in the overall length of the section, which can easily be measured. Slightly larger 'mini-satellites' with an array size varying from 0.5 to 30 kb provide the much publicized DNA fingerprinting or profiling. These sites are highly variable and abundant in human DNA. Groups can be detected simultaneously by hybridizing under conditions where differing but related sequences are probed. As each is variable the final profile will be specific for each individual. This has been accepted in most law courts and has been used for immigration and paternity cases as well as incriminating a rapist from the DNA profile left by a sperm sample. It has been proposed in the Criminal Justice Bill in the UK that a DNA sample will be extracted from everyone convicted of a serious criminal offence (Fig. 2.5).

The importance of repetitive DNA sequences in regard to recombinant DNA technology is twofold. Firstly, a gene probe which contains repetitive sequences will hybridize with many DNA fragments on a Southern blot. This will produce a smear which cannot be interpreted accurately. It is therefore essential to ensure (and there are methods for doing this) that a gene probe contains only single-copy DNA. Secondly, a probe containing a human DNA sequence belonging to the *Alu* family can be used to identify DNA of human origin in a mixture with DNA, of non-human origin, provided that the human DNA being sought contains an *Alu* sequence. This is of course very likely because of the large number of copies in the genome. Such probes have proved extremely valuable in research work on oncogenes (page 117) and also in preparing probes from particular chromosomes or chromosome regions to 'paint' cells using fluorescent *in situ* hybridization (FISH) (page 158).

TRANSCRIPTION

Genetic information contained in DNA is transmitted to messenger RNA (mRNA) by a process referred to as **transcription** which involves the DNA being used to order complementary sequences of bases in the mRNA. Genetic information in mRNA is then **translated** into protein synthesis. Much of the detail of transcription has undergone radical changes in the last few years as a result of findings revealed by recombinant DNA technology. This will be dealt with in subsequent chapters. Here we shall review only the general principles of the process.

Genes which are responsible for the synthesis of specific enzymes or peptides are referred to as **structural genes** in contrast to so-called **control genes** which modify the effects of structural genes. Enzymes and peptides are composed of amino acids, and the base composition of a structural gene codes for each of these amino acids as was first shown by Nirenberg and Matthaei in 1961 (page 2). Through the outstanding work of Brenner at Cambridge and others in the 1960s, it became clear that genetic information within the DNA molecule is stored in the form of **triplet codes**, i.e. a sequence of three nucleotide bases determines the formation of one amino acid. The triplet of bases which codes for one amino acid is called a **codon**, and the sequence of codons responsible for the synthesis of a specific polypeptide is called a **cistron**. If three bases specify one amino acid then the possible number of combinations of four bases taken three at a time where their order matters would be 4^3 or 64. But there are only twenty 'primary' amino acids commonly found in proteins. Others, e.g. hydroxyproline, are formed by the modification of one of the primary amino acids. Thus if there are 64 possible triplet codons for only twenty amino acids, some of the codons must specify more than one amino acid or have other functions. This is so. Firstly, 61 of the codons represent amino acids, all of which, except tryptophan and methionine, have more than one codon. In this regard the code is therefore sometimes said to be **degenerate**. Secondly, three codons (UAA, UAG and UGA) are responsible for the termination of protein synthesis, i.e. they are **stop codons**. Thirdly, the codon AUG nearest the 5' end of a gene is responsible for initiating protein synthesis. None of any of the other AUG codons in a gene can serve as initiation sites but instead code for methionine.

It should be noted that by convention nucleic acid sequences, whether in DNA or RNA, are always written in the direction from the 5' end to the 3' end of the molecule, and since the genetic code is actually read from the mRNA it is often represented in terms of the four bases in RNA, namely U, C, A and G. The full genetic code is given in Table 2.2.

This genetic code is universal and is found in all living organisms. The only exception is the genetic code in mitochondrial DNA which differs in detail (Table 2.3). Why there should be these differences is not known.

Table 2.2 Genetic code in terms of RNA triplets (or codons)

First base (5' end)	Second base U	Second base C	Second base A	Second base G	Third base (3' end)
U	Phenylalanine	Serine	Tyrosine	Cysteine	U
	Phenylalanine	Serine	Tyrosine	Cysteine	C
	Leucine	Serine	Stop	Stop	A
	Leucine	Serine	Stop	Tryptophan	G
C	Leucine	Proline	Histidine	Arginine	U
	Leucine	Proline	Histidine	Arginine	C
	Leucine	Proline	Glutamine	Arginine	A
	Leucine	Proline	Glutamine	Arginine	G
A	Isoleucine	Threonine	Asparagine	Serine	U
	Isoleucine	Threonine	Asparagine	Serine	C
	Isoleucine	Threonine	Lysine	Arginine	A
	Methionine	Threonine	Lysine	Arginine	G
G	Valine	Alanine	Aspartic acid	Glycine	U
	Valine	Alanine	Aspartic acid	Glycine	C
	Valine	Alanine	Glutamic acid	Glycine	A
	Valine	Alanine	Glutamic acid	Glycine	G

Differences between the universal genetic code and the genetic code in two mitochondria

Codon	Universal code	Mitochondrial code	
		Yeast	Mammalian
UGA	STOP	Tryptophan	Tryptophan
AUA	Isoleucine	Methionine	Methionine
CUA	Leucine	Threonine	Leucine
AGA AGG	Arginine	Arginine	STOP

In principle the base sequences in nucleic acids could be translated in any one of three different reading frames, as they are called, depending on where the decoding process begins. For instance the sequence:

UAAGCAUAGAU

could be read as:

UAA, GCA, UAG,

or as:

AAG, CAU, AGA,

or as:

AGC, AUA, GAU.

However, the genetic code is not overlapping in the majority of organisms including all eukaryotes. The reading frame is set at the initiation codon and proceeds sequentially thereafter, continuing until the stop codon is reached.

A further problem which puzzled molecular biologists until fairly recently is that any region of the DNA molecule could in theory be transcribed into two different mRNAs, one from each of the DNA strands. However, it now seems clear that usually only one DNA strand is copied, and the particular strand used varies throughout the genome and is gene specific. Some fascinating examples have been found where a second gene is encoded within another gene in the non-coding intervening sequences (see page 63) of the first gene. This has been observed within the factor VIII gene, gaba $\beta3$ receptor gene and the gene causing neurofibromatosis. Transcription may be in the opposite direction to the main gene. The indication as to which DNA strand is to be copied is dictated by the orientation of what is called a **promoter** sequence. Details of promoter sequences will be discussed in detail later; suffice it to say that the orientation of the promoter sequence sets off the RNA polymerase (RNA-synthesizing enzyme) in a particular direction in any given genetic region and thereby automatically determines which of the two DNA strands will be read. Since the enzyme RNA polymerase sequentially adds ribonucleoside monophosphates to the growing

3'-OH end of the RNA chain, the latter grows in the 5' to 3' direction, and the DNA strand serving as template is therefore traversed from its 3' to its 5' end. This means that each RNA molecule will be identical in its polarity and nucleotide sequence (except for the substitution of U for T) to the DNA strand that is *not* transcribed. Thus:

DNA	5'	....	TTA	CGC	GTT	ATA	....	3'
	3'	....	AAT	GCG	CAA	TAT	....	5'
RNA	5'	....	UUA	CGC	GUU	AUA	....	3'

Direction of transcription →

Though all **functioning** genes are transcribed in this way, most genomic DNA appears not to be transcribed at all. The total amount of nuclear DNA in the human haploid (gametic) genome of man amounts to 3×10^6 kb. Since in general a gene is about 1 to 20 kb in length, there should be over a million genes but in fact only about 3000 disease loci have been identified. Even taking into account all the other genes concerned with normal traits, such as height and intelligence, a large proportion of genomic DNA remains for which at present there is no defined function. This has been variously referred to as 'junk' or **selfish** DNA, the implication being that by mechanisms we do not really understand, it preserves its own existence within the genome! However, as research progresses it seems likely that more and more of this DNA will be found to have important functions, including perhaps the control of gene action.

TRANSLATION

The process whereby genetic information present in an mRNA molecule directs protein synthesis is referred to as translation. After migrating out of the nucleus into the cytoplasm, mRNA becomes associated with the ribosomes which are the site of protein synthesis. A group of ribosomes associated with the same molecule of mRNA is referred to as a **polysome**. In the ribosomes the mRNA forms the template for the sequence of particular amino acids and is therefore sometimes called template RNA. In the cytoplasm there is also another family of RNAs called transfer RNA or tRNA. Each tRNA is about 70–90 nucleotides in length and is single-stranded, but because of base pairing **within** the molecule it adopts a shape like a clover-leaf. Within the middle loop of the clover-leaf, a triplet of varying sequence forms the anticodon which can base-pair with a complementary triplet in an mRNA molecule. At the other end of the tRNA molecule is a sequence for attachment to a specific amino acid. Thus a particular triplet on the mRNA is related through tRNA to a specific amino acid. The ribosome first binds to a specific site on the mRNA molecule which thus sets the reading frame. The ribosome then moves along the mRNA molecule in a zipper-like fashion, translating one codon at a time using tRNA molecules to add amino acids to the growing end of the polypeptide chain (Fig. 2.6).

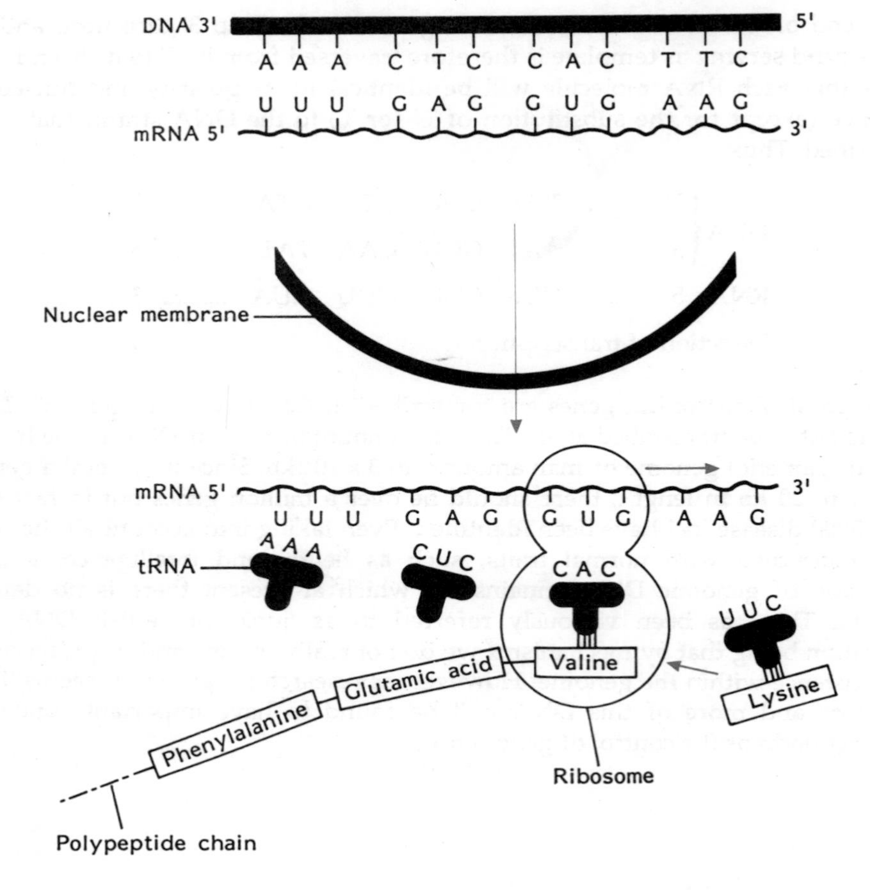

Figure 2.6 Diagrammatic representation of the way genetic information is translated into protein synthesis.

Post-translation modification

This has been a very much simplified description of an extremely complex process, the precise details of which are continually emerging. However, of particular importance to recombinant DNA technology is the so-called **post-translational modification** of proteins. This is especially important when considering the use of microorganisms for the biosynthesis of human proteins, for though microorganisms may be engineered in such a way that effective translation takes place, intracellular mechanisms for the production of an active product may be lacking.

Many proteins undergo some form of modification when they are released from the ribosome, and over a hundred different modifications are now known which may increase or decrease the functional activity of a particular protein or determine its destination in the cell. These modifications may be reversible as in the case of phosphorylation (brought about by the enzyme protein kinase), adenylation and methylation. Other modifications are permanent and are required for activity, e.g. the attachment of a coenzyme (e.g. biotin) or the cleavage by a protease of a larger protein into smaller active proteins as occurs in the production of certain hormones such as somatostatin, insulin and glucagon. Large precursor proteins or polyproteins may undergo proteolytic cleavage to produce several active peptides, and different processing pathways of a precursor protein may occur in different tissues. The addition of carbohydrate side chains, glycosylation, is particularly important in determining the uptake of molecules by cells or organelles.

In vitro protein synthesis

It is sometimes necessary to study protein synthesis *in vitro*. To do this material is used which contains all the necessary protein-synthesizing machinery, such as yeast lysate, wheat germ lysate, toad (*Xenopus*) oocytes or rabbit reticulocytes. For various reasons a cell-free extract of rabbit reticulocytes has found most favour. Rabbits are first rendered anaemic so that they produce a reticulocytosis. The animals are bled and the reticulocytes removed. These are lysed and a nuclease added to remove any endogenous mRNA. When exogenous mRNA is then added protein synthesis occurs. Such a cell-free mRNA-dependent translation system can be used in several ways. If the protein product of a particular mRNA is known, then by characterizing the protein produced by the system, the technique can be used to detect whether a particular mRNA is present or not. For example, if globin chain synthesis occurs we would know that globin mRNA was present. It may also be used as a means of identifying the basic metabolic defect in a particular disorder where this is not known. Here one uses the system to compare the products of protein synthesis using mRNAs from normal and diseased tissues.

Transgenic mice

Frequently, the final court of appeal in deciding the function of a newly isolated gene is to create a transgenic mouse and find in what way its characteristics (phenotype) are changed. Transgenic mice can be formed by injecting DNA into a fertilized egg and reimplanting it into a surrogate mother. Hopefully, if a litter of mice is born, some of them will have a fragment of the injected DNA in some of their cells, including a germ cell. If so, then it is possible to breed mice who have inserted DNA in every cell. This technique is particularly effective in examining controlling sequences around the gene: correct expression will only be found if the surrounding regulating sequences have also been transferred. It is not possible by this method to control either the number of copies inserted or the site of insertion. However, gene targeting can be achieved by homologous recombination of embryonic stem (ES) cells. The ES cell is a pluripotent, embryo derived stem cell. They are transfected with a targeting vector, introduced into the cell by electroporation or microinjection. In most cells the targeting vector inserts randomly into the ES genome but in a few cells there is insertion via homologous recombination and the mutation is introduced at the target, cognate, site. Selection or screening is necessary to identify the rare targeted cell. This is achieved by cloning of the transfected ES cells and allows the selection of appropriate clones

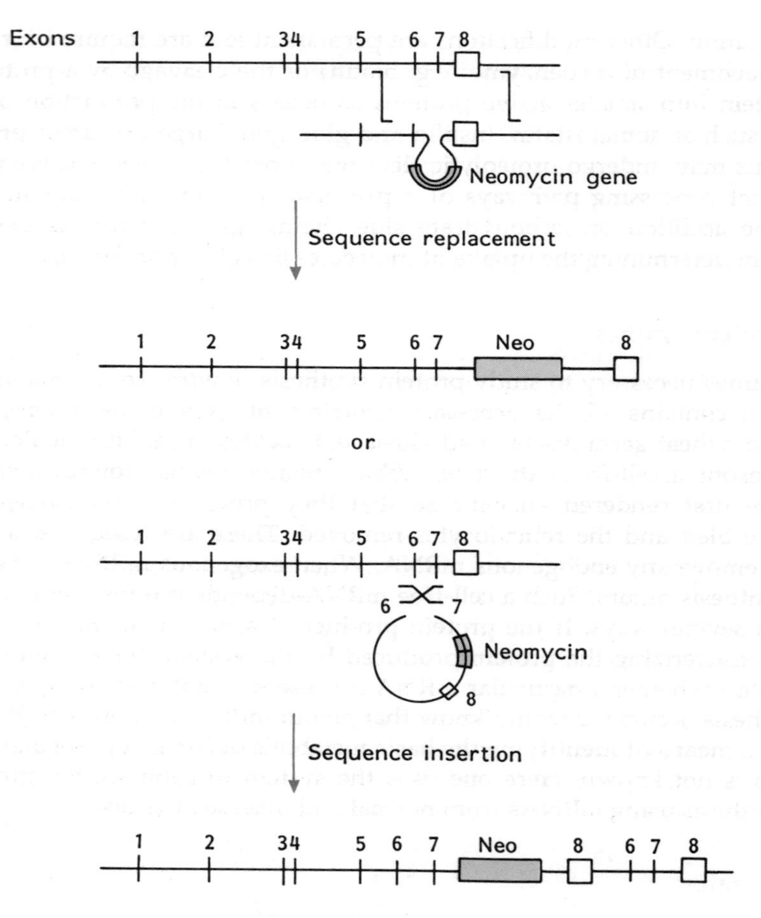

Figure 2.7 Method of targeting embryonic stem (ES) cells by homologous recombination. After homologous pairing between the same sequences in the vector and the chromosome, the original sequence is either replaced (top) or an additional sequence inserted (bottom). (Adapted from Capecchi, 1989.)

which are then injected into a blastocyst which, as above, is implanted into a foster mother. The key to the homologous recombination is the similarity of the sequences in the vector and host. Both sequence replacement and sequence insertion vectors can be designed (Fig. 2.7). Various 'knock-out' mice, with a gene specifically disrupted, have been produced this way, which makes it possible to trace the development and natural history of an animal without the particular gene.

MUTATIONS

A mutation may be defined as a change in the genomic material and may, occur in somatic cells or in the germ-line (gametes). Only gametic mutations are inherited and are important because they are the cause of genetic variability and the basis for natural selection and evolution. However, until relatively recently, not a great deal was known about the molecular basis of mutation. This is therefore perhaps a convenient point to consider the matter because an understanding of the nature of mutations is very important in order to appreciate how various DNA techniques are used, for example, in the prenatal diagnosis of genetic disease.

Mutations can be considered under two main headings: large chromosomal rearrangements and nucleotide base changes. The former include inversions, duplications and deletions of large segments of chromosomal DNA. Base changes, or **point mutations**, on the other hand, are submicroscopic and are the commonest type of mutation, and will be our main concern here. They may involve either the deletion or insertion of additional bases, or the substitution of one base for another. The deletion or insertion of one or two bases will result in a complete change in the amino acid sequence of a protein 'downstream' from the mutation because of the change in the reading frame for all subsequent codons in the gene. These are therefore referred to as **frameshift mutations**. If three (or a multiple of three) bases are eliminated the reading frame will not be changed and the effects may be less drastic because only one or a few amino acids will be missing in the protein product. Acridine compounds have the property of distorting the DNA molecule in such a way that additional bases may be incorporated and others deleted. These substances therefore specifically produce frameshift mutations (Fig. 2.8).

Mutations associated with base substitutions are of two different types: **transitions**, where a purine is replaced by another purine or a pyrimidine by another pyrimidine, and **transversion**, where a purine is replaced by a pyrimidine and vice versa. Base substitutions may occur spontaneously because by chance a wrong base becomes incorporated in the DNA molecule during replication. This is called a **copying error**. Such errors may also be induced by mutagens such as hydroxylamine. This reacts with cytosine to produce hydroxylated derivatives which can pair with adenine instead of guanine. In the next cell division the adenine which paired with the hydroxylated cytosine now pairs with the usual thymine. Therefore a C–G pair is replaced by a T–A pair, i.e. a transition type of mutation has been produced. Some compounds, such as nitrosamines, preferentially (but not exclusively) produce transitions whereas others, such as benzpyrene and aflotoxin, tend to produce transversions. It is interesting to note that these various mutagenic agents are also powerful carcinogens.

A base substitution may have one of several effects. It may result in a change of an amino-acid-specifying codon, and is then referred to as a **missense mutation**.

Normal:	Paula is our president
Substitution:	Paula id our president
Deletion:	Paula is president
Frameshift:	Paula so urp resident

Figure 2.8 **Mutations.**

This may have no detectable effect in the organism and is only recognized by DNA sequencing. Such 'silent' mutations may occur either because the mutation results in the substitution of the same amino acid (because most amino acids are specified by more than one codon) or the substitution of a different amino acid which does not affect the biological activity of the gene product. However, a missense mutation may result in the substitution of an amino acid which does affect the activity of the protein product, resulting usually in no activity or reduced activity, or rarely in increased activity. Missense mutations are also known which do not themselves directly affect the biological activity of the protein product but may do so indirectly by affecting its molecular shape or by rendering it more vulnerable to catalytic breakdown. A base substitution may also result in the generation of a new stop codon within the gene itself, producing premature termination of translation and the shorter protein product may or may not have normal activity. This is sometimes referred to as a **nonsense mutation**. Finally, a mutation may eliminate the normal stop codon with the effect that translation then continues until another stop codon beyond the normal gene is reached and the protein produced will therefore be abnormally elongated.

It was observed some time ago that the dinucleotide sequence 5'CpG3' is considerably under-represented in DNA compared to statistical expectations. A C in this position is the major site of methylation in human DNA. ^{me}C can become deaminated to T but as T occurs naturally in human DNA the systems which check for errors during replication fail to detect this. In further cycles of replication an A is incorporated opposite the T, replacing the original G in the opposite strand. 5'CpG3' to 5'TpG3' is indeed found commonly as a point mutation in human genes, up to 50% of all mutations in some genes, causing something of mutation hot spots and this is believed to be the mechanism by which it arises.

So far we have only been considering mutations in the DNA molecule. These are certainly the most important but mutations may also occur in tRNA anticodons and have been extensively studied in lower organisms. They are recognized because they overcome the effects of mutations of genes coding for proteins and are therefore referred to as **suppressor mutations**. For example, the mutation of the DNA codon GGA to AGA will result in the replacement of glycine by arginine. However, if the tRNA anticodon CCU for glycine were also to mutate to UCU, glycine would continue to be incorporated in the protein product despite the mutation in the DNA codon.

A conceivable application of such suppressor mutations might one day be in gene therapy. It is now possible, from work pioneered in the Republic of China, to synthesize biologically active tRNAs. Into such synthetic tRNAs specific suppressor mutations could perhaps be incorporated which might suppress disease-producing mutations.

All mutations, whether they occur spontaneously or are induced in some way, are random events. However, technology is now becoming available by which it is possible to produce specific mutations at precise sites; this has therefore been called **site directed mutagenesis**. Mutations are constructed, with novel regulating signals or different codons at predetermined sites, in cloned genes in order to define precisely their effects on gene function. In this way more is being learnt about the function of promoter sequences and the details of gene transcription and translation. The original scheme was worked out by Smith, for which he received the Nobel prize in 1993.

Vectors have been adapted so that the introduction of chosen mutations into the coding sequence of a protein is now a straightforward procedure. This uses synthetic oligonucleotides and a phagemid, a derivative of the single-stranded M13 phage. The DNA sequence of interest is introduced into the phage by the usual methods of DNA technology. An oligonucleotide corresponding to a part of the DNA sequence of interest is synthesized chemically but includes a particular mutation. The synthetic oligonucleotide hybridizes to the homologous part of the DNA sequence of interest and acts as a primer to synthesize a double-stranded duplex in the presence of DNA polymerase. This hybrid phagemid is then used to transform *E. coli* or a similar host. At the first round of replication the mutant sequence segregates from the normal. The mutant is then recovered by various

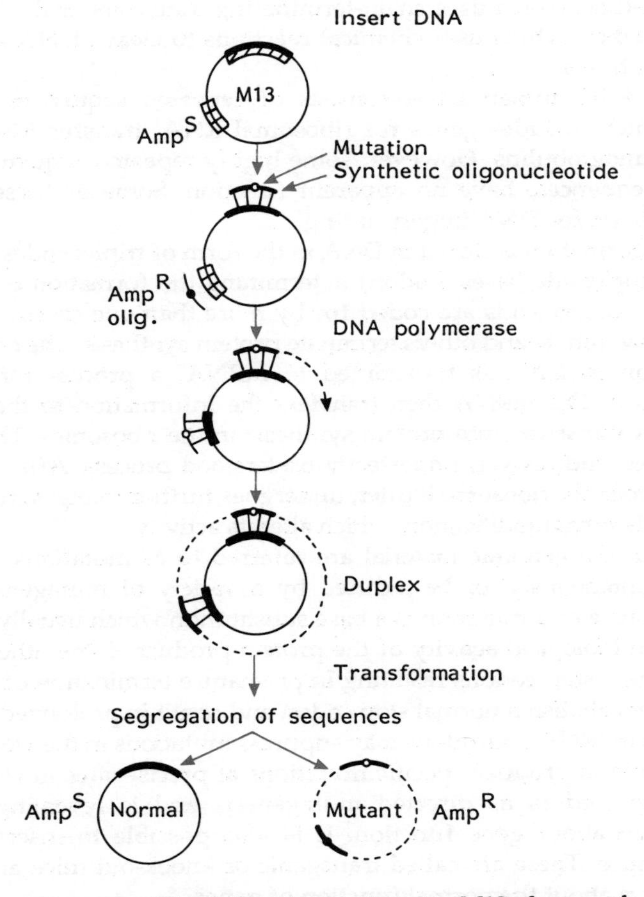

Figure 2.9 Directed mutagenesis using M13 phage and a synthetic oligonucleotide.

screening methods and its expression compared with the normal (Fig. 2.9). On average half of the molecules should contain the base change but in practice the efficiency is much lower than that.

The whole system is made much more efficient by also mutating a mutant inactive gene for ampicillin resistance back to its active form. Only colonies which are now ampicillin resistant are likely to have been mutated.

There is no denying the technical complexities of these various approaches to directed mutagenesis, but they are likely to produce important information about gene function which cannot be easily obtained in any other way.

SUMMARY

- The DNA molecule is composed of two polynucleotide chains arranged in a double helix. The sequence of nucleotide bases may be determined by the Sanger method which uses chain-terminating inhibitors or the Maxam and Gilbert method which uses chemical reactions to cleave DNA at particular nucleotide bases.
- About 40% of human DNA consists of repeated sequences of varying length which includes genes for ribosomal RNA, transfer RNA, histones and immunoglobulins. However, some highly repeated sequences, such as the *Alu* sequences, have no apparent function. Some of these sequences form the basis for DNA fingerprinting.
- Genetic information is stored in DNA in the form of triplet codes, a sequence of three nucleotide bases (codon) determining the formation of one amino acid. Most amino acids are coded for by more than one codon, and certain codons may initiate and others terminate protein synthesis. The coded genetic information in DNA is transmitted to mRNA, a process referred to as transcription. The mRNA then transfers the information to the cytoplasm where it is translated into protein synthesis in the ribosomes. The latter is a complicated and, as yet, imperfectly understood process. After a protein is released from the ribosome it often undergoes further changes, referred to as post-translational modification, which alter its activity.
- Changes in the genomic material are referred to as mutations, which may occur 'spontaneously' or be induced by a variety of mutagens. The most frequent cause of a mutation is a base substitution which usually results in a decrease in biological activity of the protein product. Some other mutations generate new stop codons resulting in premature termination of translation, while others abolish a normal stop codon and result in prolonged translation. Mutations in tRNA anticodons may suppress mutations in the DNA.
- It is possible to produce specific mutations at precise sites in cloned genes. This is referred to as directed mutagenesis and is revealing important information about gene function. It is also possible to insert or mutate genes in mice. These are called transgenic or knock-out mice and also give information about the normal function of genes.

REFERENCES AND FURTHER READING

Textbooks and review articles

Bauer WR, Crick FHC, White JH. Supercoiled DNA. *Sci Amer* 1980; **243**(1): 100–113

Calladine CR, Drew HR. *Understanding DNA: The Molecule and How it Works*. London: Academic Press, 1992

Collins FS, Galas D. A new five-year plan for the U.S. Human Genome Project. *Science* 1993; **262**: 43–46

Crick, FHC. The genetic code. *Sci Amer* 1966; **215**(4): 55–62

Kornberg RD, Klug A. The nucleosome. *Sci Amer* 1981; **244**(2): 48–60

Lewin B. *Genes* V. Oxford: OUP, 1994

McKusick VA. *Mendelian Inheritance in Man*, 10th edn. London and Baltimore: Johns Hopkins Press, 1992

Watson JD, Hopkins NH, Roberts JW, Steitz JA, Weiner AM. *Molecular Biology of the Gene*, 4th edn. California: Benjamin/Cummings, 1987

Watson JD, Gilman M. *Recombinant DNA*, 2nd edn. New York: Freeman, 1992

Research publications

Anderson S, Banker AT, Barrell BG, de Bruijn MHL, Coulson AR, *et al*. Sequence and organization of the human mitochondrial genome. *Nature* 1981; **290**: 457–465

Capecchi MR. Altering the genome by homologous recombination. *Science* 1989; **240**: 1288–1292

Jeffreys AJ, Wilson V, Thein SL. Hypervariable 'mini-satellite' regions in human DNA. *Nature* 1985; **314**: 67–73

Jeffreys AJ, Wilson V, Thein SL. Individual specific 'fingerprints' of human DNA. *Nature* 1985; **316**: 76–79

Maxam AM, Gilbert WA. A new method for sequencing DNA. *Proc Natl Acad Sci USA* 1977; **74**: 560–564

Sanger F. Determination of nucleotide sequences in DNA. *Science* 1981; **214**: 1205–1210

Sanger F, Nicklen S, Coulson AR. DNA sequencing with chain-terminating inhibitors. *Proc Natl Acad Sci USA* 1977; **74**: 5463–5467

Steiner DF, Quinn PS, Chan SJ, Marsh J, Tager HS. Processing mechanisms in the biosynthesis of proteins. *Ann NY Acad Sci* 1980; **343**: 1–16

Wang DB, Zheng KQ, Qui MS, Liang ZH, Wu RL, *et al*. Total synthesis of yeast alanine transfer RNA. *Scientia Sinica* 1983; **26**: 464–481

Chapter 3
The technology

DNA technology has introduced into the biological sciences a precision and emphasis on meticulous detail which in the past had been largely the province of the physical sciences. Some knowledge of at least the basic principles of this technology is necessary in order to appreciate developments in the field, and these principles will be discussed in this chapter. For details of the various practical procedures the original publications can be consulted, and several laboratory manuals of recombinant DNA technology are available.

The actual recombinant part of the technology, whereby new combinations of genetic material are formed, is relatively well defined and will be dealt with here. However, the polymerase chain reaction (PCR) has rapidly become an essential technology often, but by no means always, replacing the need to clone a sequence. It is convenient to consider the recombinant part of the technology under four main headings: firstly, the generation of DNA fragments using, for example, restriction endonucleases; secondly, the incorporation of these fragments into a suitable vector; thirdly, the introduction of the vector into a particular host organism which is then grown in culture to produce clones with multiple copies of an incorporated DNA fragment; finally, the selection and harvesting of clones which contain a specific DNA fragment.

RESTRICTION ENDONUCLEASES

DNA may be reduced to fragments by mechanical shearing. This is not very helpful because the fragments so produced will be of varying sizes and since the process is random there is no way of ensuring that a particular DNA sequence is isolated. For this reason recombinant DNA technology had to await the discovery of a group of enzymes, the restriction endonucleases, which could accomplish this. Some of these cleave DNA at random, but here we shall only be concerned with those (so-called type II) which cleave DNA at sequence-specific sites. It was this specificity which proved of immeasurable value in DNA technology.

These enzymes are found naturally largely in bacteria and are referred to as **restriction endonucleases**. They are endonucleases because they cleave DNA within (as opposed to the ends of) the molecule and are called **restriction** endonucleases because they restrict their activity to 'foreign' DNA. If DNA from one strain of *E. coli* is introduced into a different strain the former is fragmented by restriction endonucleases possessed by the latter and loses its function. The host's DNA is not so attacked because the sites vulnerable to its own enzymes are

protected by a process of methylation. In fact a bacterial cell has two types of site-specific enzymes: one methylates and thus protects specific sites, while the other cleaves these sites if they are not protected. In nature restriction endonucleases act as a means by which a particular bacterial strain protects itself against violation by 'foreign' DNA. It is intriguing to speculate that if such a system had not been

Table 3.1 Some restriction endonucleases and their recognition sequences and cleavage sites (*). Py = pyrimidine, Pu = purine, N = any base. Note that some enzymes (isoschizomers) recognize the same sequences (e.g. *Hind*III and *Hsu*I; *Hpa*II and *Msp*I). [ME] = Methylation sensitive

Enzyme	Organism	Cleavage site (*) 5' 3'
Tetranucleotides (4)		
*Alu*I	*Arthrobacter luteus*	A G * C T
*Hae*III	*Haemophilus aegyptius*	G G * C C
*Hpa*II	*Haemophilus parainfluenzae*	C * C G G
*Mbo*I	*Moraxella bovis*	* G A T C
[ME] *Sau*3A	*Staphylococcus aureus* 3A	* G A T C *
*Taq*I	*Thermus aquaticus*	T * C G A
*Msp*I	*Moraxella species*	C * C G G
Pantanucleotides (5)		
*Ava*II	*Anabaena variabilis*	C * G A/T C C
*Dde*I	*Desulfovibrio desulfuricans*	C * T N A G
*Eco*RII	*Escherichia coli* R.245	* C C A/T G G
*Hin*fI	*Haemophilus influenzae* Rf	G * A N T C
Hexanucleotides (6)		
[ME] *Ava*I	*Anabaena variabilis*	C * Py C G Pu G
*Bal*I	*Brevibacterium albidum*	T G G * C C A
*Bam*HI	*Bacillus amyloliquefaciens* H	G * G A T C C
*Bgl*II	*Bacillus globigii*	A * G A T C T
*Eco*RI	*Escherichia coli* RY13	G * A A T T C
*Hae*II	*Haemophilus aegyptius*	Pu G C G C * Py
*Hinc*II	*Haemophilus influenzae* Rc	G T Py * Pu A C
*Hind*III	*Haemophilus influenzae* Rd	A * A G C T T
*Hpa*I	*Haemophilus parainfluenzae*	G T T * A A C
*Hsu*I	*Haemophilus suis*	A * A G C T T
*Pst*I	*Providencia stuartii*	C T G C A * G
*Sac*I	*Streptomyces achromogenes*	G A G C T * C
[ME] *Sal*I	*Streptomyces albus*	G * T C G A C
[ME] *Sma*I	*Serratia marcescens*	C C C * G G G
[ME] *Xma*I	*Xanthomonas malvacearum*	C * C C G G G
*Sph*I	*Streptomyces phaeochromogenes*	G C A T G * C
Heptanucleotides (7)		
*Mst*II	*Microcoleus species*	C C * T N A G G
Octanucleotides (8)		
[ME] *Not*I	*Nocardia otitidis-caviarum*	G C * G G C C * G C

present in microorganisms' evolution as we imagine it might not have even commenced!

Each enzyme is designated according to the organism from which it is derived: the first initial of the genus together with the first two initials of the species of the particular microorganism. These three letters may then be followed by a strain designation if the enzyme is present in a specific strain. Finally, a Roman numeral is used to indicate the order of discovery of an enzyme in a particular strain. For example, *Eco*RI is an enzyme obtained from *Escherichia coli* strain R and was the first such enzyme isolated.

Some enzymes recognize sequences four nucleotides long, others five nucleotides long, but most six nucleotides long (Table 3.1). Very few recognize longer sequences. Some enzymes termed isoschizomers may cleave the same site but be obtained from different organisms (e.g. *Hin*dIII and *Hsu*I) or recognize the same sequence but cleave at different sites within the sequence (*Xma*I and *Sma*I). Some enzymes recognize different sequences but with the same inner sequences (*Bam*HI and *Sau*3A). Enzymes which have a high proportion of Cs and Gs in their site cut relatively infrequently and, therefore, produce very long pieces of DNA. These have proved very useful in gene mapping as they can show that two sequences are adjacent within the genome. So-called HTF (*Hpa*II tiny fragments) islands are CG rich and associated with the 5' ends of unmethylated (active) genes. The search for such islands (with appropriate probes) in gene libraries can be used to search for genes, as was the case in cystic fibrosis. By convention the recognition sequences are written in the 5' to 3' direction and are only given for one DNA strand.

Many hundreds of restriction enzymes with different specificities have now been isolated. Commercial companies are constantly updating their range and introducing cheaper isoschizomers. Most are obtained from various strains of bacteria but not all. A few have been isolated from other microorganisms, and some of these have unique and demanding growth requirements. For example, the enzyme *Aha*III, which cuts TTT*AAA, has been isolated from the alga *Aphanothece halophytica* which will only grow in very high salt concentrations at a temperature of 60–65 °C! Note, that *Taq*I from *Thermophilus aquaticus* works at 65 °C, as may be surmised from the name; other enzymes from this organism, e.g. DNA polymerase, also operate at an elevated temperature.

Because of complementary base pairing in the DNA molecule, restriction endonucleases create double-stranded breaks. Thus the enzyme *Eco*RI specifically cleaves between G and A in the sequence – G A A T T C – and thus produces staggered ends:

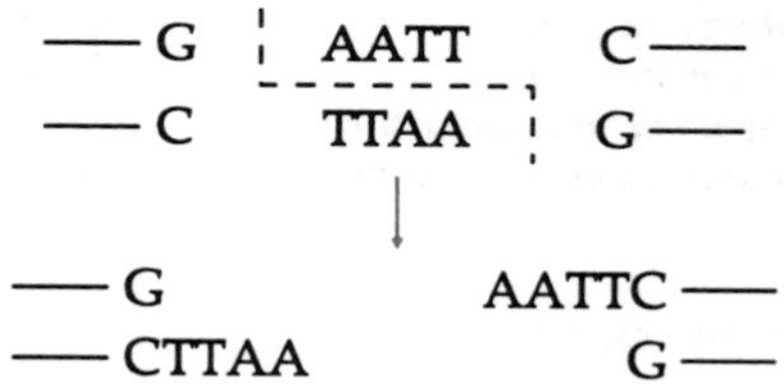

```
—G ¦ AATT    C—
    -------¦
—C   TTAA ¦ G—
        ↓
—G             AATTC—
—CTTAA             G—
```

The staggered ends with complementary bases are called 'sticky' because they will combine with similar sequences produced by the same enzyme on the DNA of a suitable vector. Other enzymes, however, produce blunt-ended fragments, e.g. *Sma*I:

```
— CCC ¦ GGG —
— GGG ¦ CCC —
      ↓
— CCC     GGG —
— GGG     CCC —
```

*Sma*I is derived from *Serratia marcescens*, a Gram negative bacillus which characteristically produces a bright red pigment, and it has been suggested it may have been responsible for some of the stories from the Middle Ages of communion bread mysteriously dripping 'blood'!

The different types of ends produced by restriction enzymes have important implications in the strategy subsequently used to join two DNA molecules together.

Joining DNA molecules

When a restriction enzyme produces staggered sticky (cohesive) ends, as in the case of *Eco*RI, and is used on both the DNA to be incorporated into a vector ('foreign' DNA) as well as on the DNA of the vector itself, the cohesive ends will come together and be held by hydrogen bonding between the complementary bases, a process sometimes referred to as annealing. The two molecules can then be sealed (ligated) and stabilized by the joining enzyme **DNA ligase**. There are several types of ligases and some (e.g. T4 DNA ligase) will actually link together DNA molecules with blunt ends. However, such blunt-ended ligation is much less efficient than cohesive-ended ligation, and when blunt ends are produced by a restriction enzyme then other strategies of joining the DNA molecules together have to be considered. One way is to add to the ends of a DNA molecule a synthetic DNA linker which is designed to contain sequences recognized by a restriction enzyme

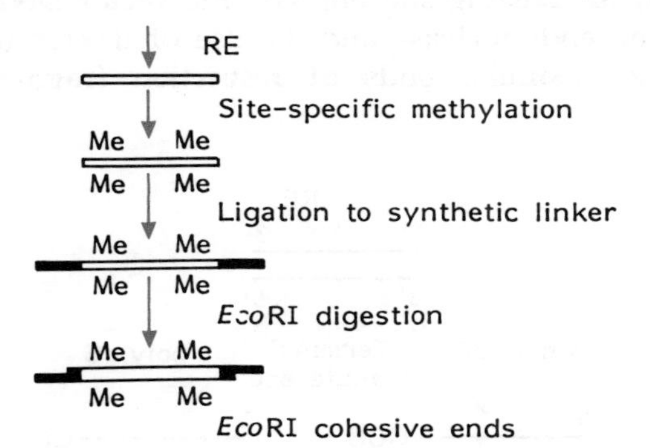

Figure 3.1 The use of synthetic DNA linker molecules where an initial restriction endonuclease (RE) produces blunt ends. 'Me' indicates *Eco*RI sites which have been 'protected' by methylation.

which will produce staggered cohesive ends. In this case it may be necessary to first protect the DNA itself from attack by the enzyme by methylating the sites which will be cleaved by the enzyme. The technique is illustrated in Fig. 3.1.

Another approach is to add complementary bases to the cut ends of both the foreign DNA and the vector DNA. The enzyme **terminal transferase** specifically adds bases (deoxynucleotides) to the 3' ends of a DNA molecule and by using this enzyme it is possible to add so-called **homopolymer tails**, i.e. end sequences containing only one particular base (dG, dC, dA or dT). Thus poly dG (GGG ...) may be added to the 3' ends of the foreign DNA and poly dC (CCC ...) to the 3' ends of the vector DNA. When the two are mixed complementary base pairing will occur and the DNA molecules will join together and can then be ligated (Fig. 3.2).

These methods – cohesive ligation, synthetic DNA linkers and homopolymer tailing – form the basis for joining DNA molecules but do not exhaust the possibilities. A variety of enzymes may be used in these techniques, some of which are given in Table 3.2. Under certain circumstances the generation of blunt-ended fragments may have an advantage and the enzyme S1 nuclease, which degrades single-stranded DNA, can be used to trim a staggered end to produce a blunt end, or the shorter strand of a staggered end may be extended (filled in) by adding nucleotides using DNA polymerase. Occasionally even different enzymes can be used on the foreign and vector DNAs because a certain degree of mismatching of the end sequences may be corrected subsequently by various techniques referred to as filling-in and resealing (see page 34). It is even possible to reconstruct **new** restriction sites at a junctional region. However, details of these methods are outside the scope of this book but emphasize the usefulness of a sound biochemical background in carrying out these techniques.

The choice of a particular restriction endonuclease and DNA-joining strategy depends on a number of factors. In general preference is given to an enzyme where there is only a single site on both the foreign and vector DNAs and which is readily available and relatively inexpensive. Otherwise it may be necessary to resort to a variety of ingenious techniques in order to generate, for example, **new** restriction sites or even destroy an existing site on, say, the vector DNA. The choice of an appropriate restriction endonuclease and the use of linkers (and single-stranded **adaptors** for fusing dissimilar ends of restriction fragments) are extremely

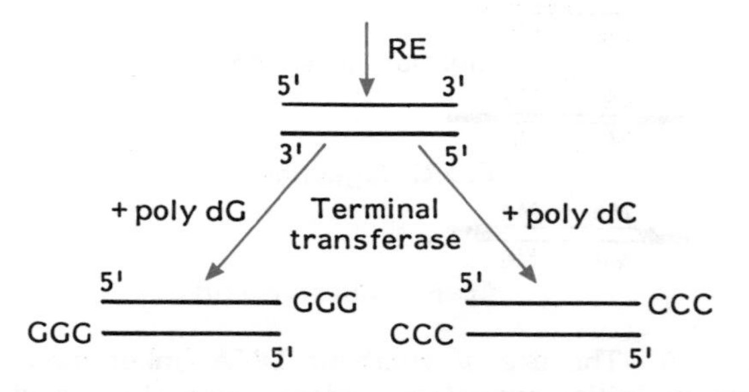

Figure 3.2 The use of homopolymer tailing where a restriction endonuclease (RE) produces blunt ends.

Table 3.2 Enzymes used in DNA technology

Enzyme	Function	Main uses
Alkaline phosphatase	Dephosphorylates 5' ends of RNA and DNA	Prevention of self-ligation
DNA ligase	Catalyses bonds between DNA molecules	Joining DNA molecules
DNA polymerase (e.g. 'Klenow fragment')	Synthesizes double-stranded from single-stranded DNA	(a) Synthesis of double-stranded cDNA (b) Nick translation
DNase I	Under appropriate conditions produces single-stranded nicks in DNA	Nick translation
Exonuclease III	Removes nucleotides from 3' ends of DNA	DNA sequencing
λ Exonuclease	Removes nucleotides from 5' ends of DNA	DNA sequencing
Nuclease *Bal*31	Degrades both the 3' and 5' ends of DNA	Progressive shortening of DNA molecules
Polynucleotide kinase	Transfers phosphate from ATP to DNA or RNA	^{32}P labelling of DNA or RNA
Restriction endonucleases (type II)	Cleave DNA at sequence-specific sites	Generation of recombinant DNA
Reverse transcriptase	Synthesizes DNA from RNA	Synthesis of cDNA from mRNA
SI nuclease	Degrades single-stranded DNA	Removal of 'hairpin' in synthesis of cDNA
Taq polymerase	Heat stable. Synthesizes double-stranded DNA from single-stranded DNA	Polymerase chain reaction
Terminal transferase	Adds nucleotides to the 3' ends of DNA	Homopolymer tailing

important considerations in recombinant DNA technology. Many vectors have now had a multiple cloning site added, a stretch of DNA ingeniously designed to contain many restriction sites, in order to make the choice of suitable cloning enzymes wider (Fig. 3.3). This will become clearer when we begin to discuss some of the applications of the technology.

Restriction mapping

Earlier we mentioned that DNA molecules of different sizes can be separated by electrophoresis on a medium such as agarose gel, the smaller fragments migrating faster and therefore further along the gel than larger fragments. The DNA fragments produced by the restriction enzyme may be separated in this way, and by the sequential use of different enzymes it is possible to arrange (map) the

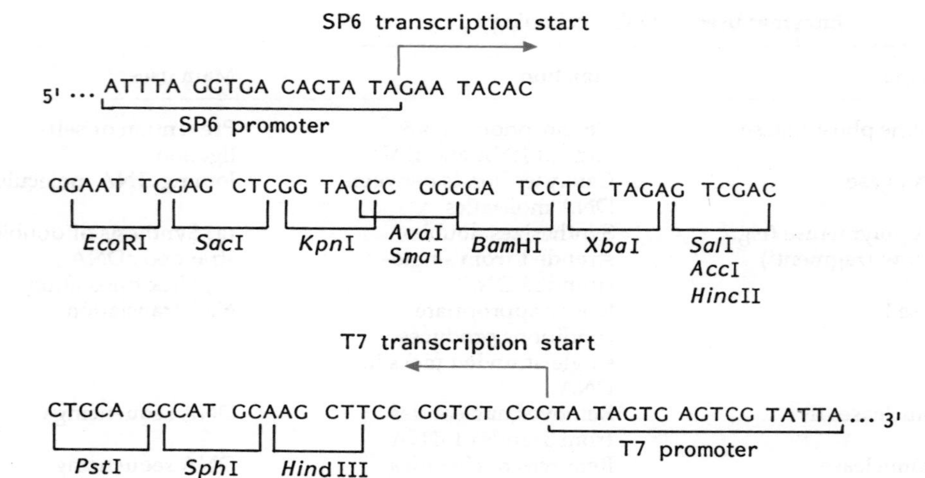

Figure 3.3 pGEM-4 plasmid promoter and multiple cloning site sequence. The wide choice of enzymes makes cloning simpler and more flexible.

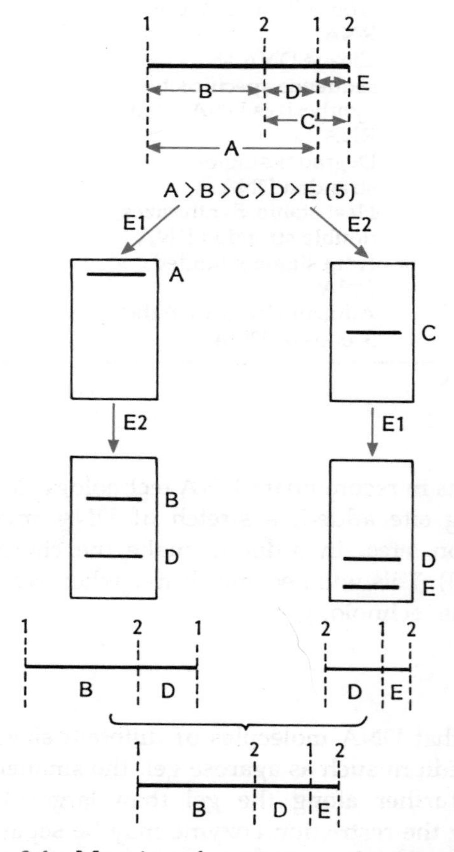

Figure 3.4 Mapping of restriction endonuclease sites.

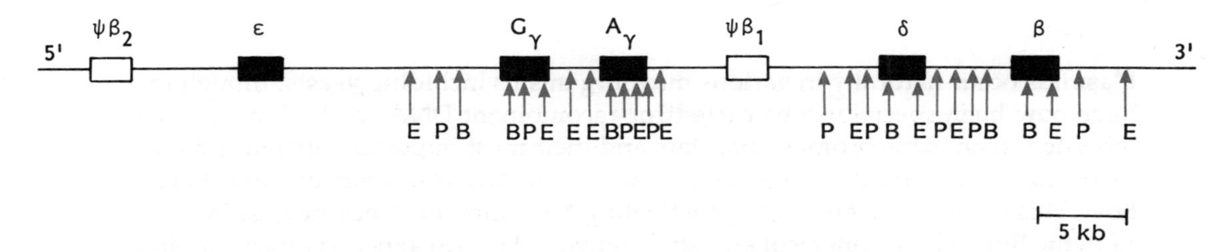

Figure 3.5 Restriction sites in the region of the human β-globin gene on chromosome 11. The black boxes indicate functional genes, the open boxes indicate pseudogenes. E = *Eco*RI; B = *Bam*HI; P = *Pst*I.

position of the sites of these enzymes. This is referred to as **restriction mapping** and is illustrated in Fig. 3.4.

If enzyme E1 cuts the DNA at site 1 and is used first, a large fragment A will be produced, and if this is then exposed to enzyme E2, which cuts the DNA at site 2, two smaller fragments (B and D) will be produced. Similarly, if enzyme E2 is used first, then a large fragment C will be produced, and if this is then exposed to enzyme E1 two smaller fragments (D and E) will be produced. For illustrative purposes it is assumed the various DNA fragments are in decreasing order of size from A to E (Fig. 3.4). By knowing the order of the restriction fragments on the gel it is possible to surmise the order of the restriction sites in the original DNA. Since it is also possible to use a prepared DNA marker which produces fragments of known sizes (in kilobases) on electrophoresis, the sizes of the fragments produced by restriction enzymes can be determined and therefore the exact distances apart of the restriction sites so deduced. There are a number of variations on this theme.

The technique has important uses. If gene loci are contained in any of the DNA fragments then the technique can be used to help order these genes along the chromosome. Also knowing where the restriction sites are located may help to provide a means of cutting out (excising) a particular gene or group of genes. Some restriction sites which have been mapped around the β-globin gene region of man are illustrated in Fig. 3.5. Note the similarity of the restriction map around the two gamma globin genes (G_γ and A_γ) reflecting the strong similarity of the two sequences. The δ and β globin genes are also very similar in sequence to each other and share common restriction sites.

VECTORS

According to the *Oxford English Dictionary* the word 'vector' was first introduced into the English language in 1704 to mean a 'carrier' and was adopted by molecular biologists in the early 1970s to describe vehicles used for cloning DNA. Four types of vectors are in common use – plasmids, bacteriophage, cosmids and yeast artificial chromosomes (YACs). The first three replicate independently of their host organism and are therefore sometimes also called **replicons**.

Plasmids

Plasmids occur naturally in various microorganisms including yeasts, though many have now been specially constructed for recombinant DNA work. They are stably inherited in an extrachromosomal state and their most important attribute, as far as medicine is concerned, is that they confer antibiotic resistance on their bacterial host. This is usually achieved by inactivating the antibiotic concerned, as in the case of penicillin, chloramphenicol and streptomycin, but resistance to other antibiotics, such as tetracycline, is mediated by other plasmid-controlled mechanisms. Apart from drug resistance, plasmids may also confer resistance to heavy metals and some are involved in the degradation of various hydrocarbons. They can cause serious enteritis in piglets and possibly some cases of 'summer diarrhoea' in humans as well as crown-gall disease in plants. Some plasmids also produce proteins called colicins which kill other bacteria.

Structurally they consist of a circular duplex of DNA and possess a limited number of specific restriction enzyme sites. A plasmid frequently used for cloning in recombinant DNA technology, Bluescript, is represented diagrammatically in Fig. 3.6. It carries an antibiotic resistance gene for ampicillin so that bacteria

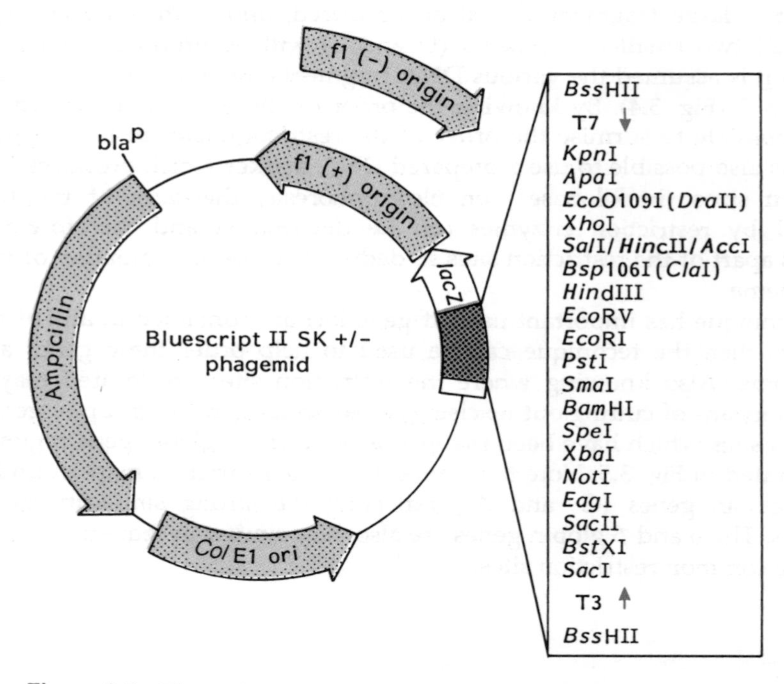

Figure 3.6 Bluescript is a convenient and commonly used plasmid for cloning. There are 21 unique restriction sites in the multiple cloning region. The plasmid is resistant to ampicillin. Bacterial promoters on either side of the cloning site allow manufacture of probes and expression of a fusion peptide. The *lacZ* gene allows colour selection of recombinants. Because the vector contains the F1 origin of replication from the F1 filamentous phage, single-stranded DNA can be recovered with a helper phage.

containing copies of it can be selected by growing in media containing ampicillin. There is a very convenient synthetic polylinker with multiple restriction enzyme sites which will make both cloning and mapping easy. The presence of the *lacZ* gene allows selection of plasmids containing cloned inserts (see page 45) and it can be produced in a single-stranded phage to make sequencing easier.

An important consideration when selecting a plasmid for DNA cloning is its **copy number**, i.e. the number of copies the plasmid produces on replication in a bacterial cell. This varies from a few to as many as 200. In general the smaller a plasmid and the greater its copy number the more useful it is as a cloning vector.

Bacteriophage

Bacteriophage (or phage for short) are viruses which infect bacteria. Each phage particle consists of a head and a tail and has a protein coat which surrounds a central core of DNA. By means of its tail it becomes attached to a bacterium and injects its DNA into the host cell. Here the phage DNA circularizes by its two ends, referred to as **cos sites** (cohesive end sites), joining together. At this stage in certain phage, called temperate phage, one of two alternative pathways of replication may be chosen. The infecting phage genome may become integrated into the circular bacterial chromosome and then replicate along with it. This is referred to as the **lysogenic** phase. Alternatively, the circular DNA of the phage may undergo replication, independently of the host DNA, to produce a large number of progeny particles which then burst (lyse) the host cell and invade other bacteria. This is referred to as the **lytic** phase. Because phage spread from one bacterial cell to another soon clear areas, or plaques, appear in a confluent layer of cultured bacteria, and the existence of such plaques indicates that bacteria have become infected with phage particles. Phage in the lysogenic phase may be induced to enter the lytic phase by a variety of external agents including exposure to ultraviolet light.

The best studied phage is the so-called lambda (λ) phage which possesses some sixty genes, all of which have now been carefully mapped, and the complete sequence of its 48 000 base pairs has been determined. Clustered in the middle of the genome (Fig. 3.7) are genes involved in the lysogenic phase of its life cycle, and since the phage must replicate in the lytic phase if it is to be used as a cloning vehicle, these can be deleted or substituted. In this way over a hundred vectors have been derived from phage λ. These include the various λ gt (generalized

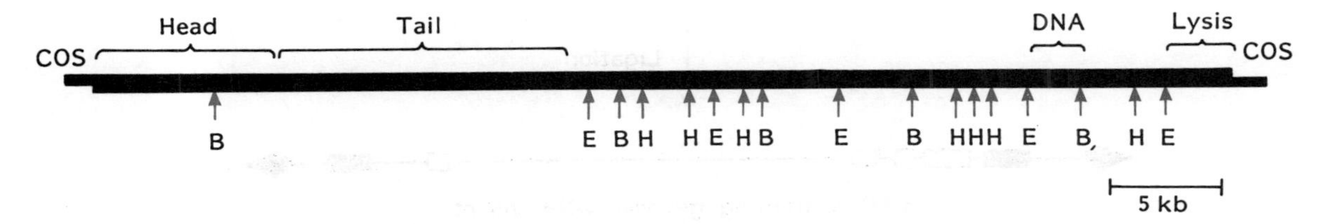

Figure 3.7 Phage lambda. The sites of three restriction enzymes are indicated as well as genes involved in the synthesis and assembly of head and tail proteins, DNA synthesis and host lysis. E = *Eco*RI; B = *Bam*HI; H = *Hin*dIII.

transducing) and Charon phage, the latter called after the character in Greek mythology who ferried the dead across the river Styx. All these vectors possess restriction sites in sequences not essential for replication and lysis and which can thus be replaced by foreign DNA for cloning.

A useful development has been the introduction of several single-stranded DNA phage as cloning vectors, the so-called M13 vectors. These have the advantages that by being single-stranded, DNA sequencing by the Sanger method is suitable (page 17) and cloning DNA to be sequenced in M13 provides an ideal template for the introduction and study of defined site-directed mutations (page 29).

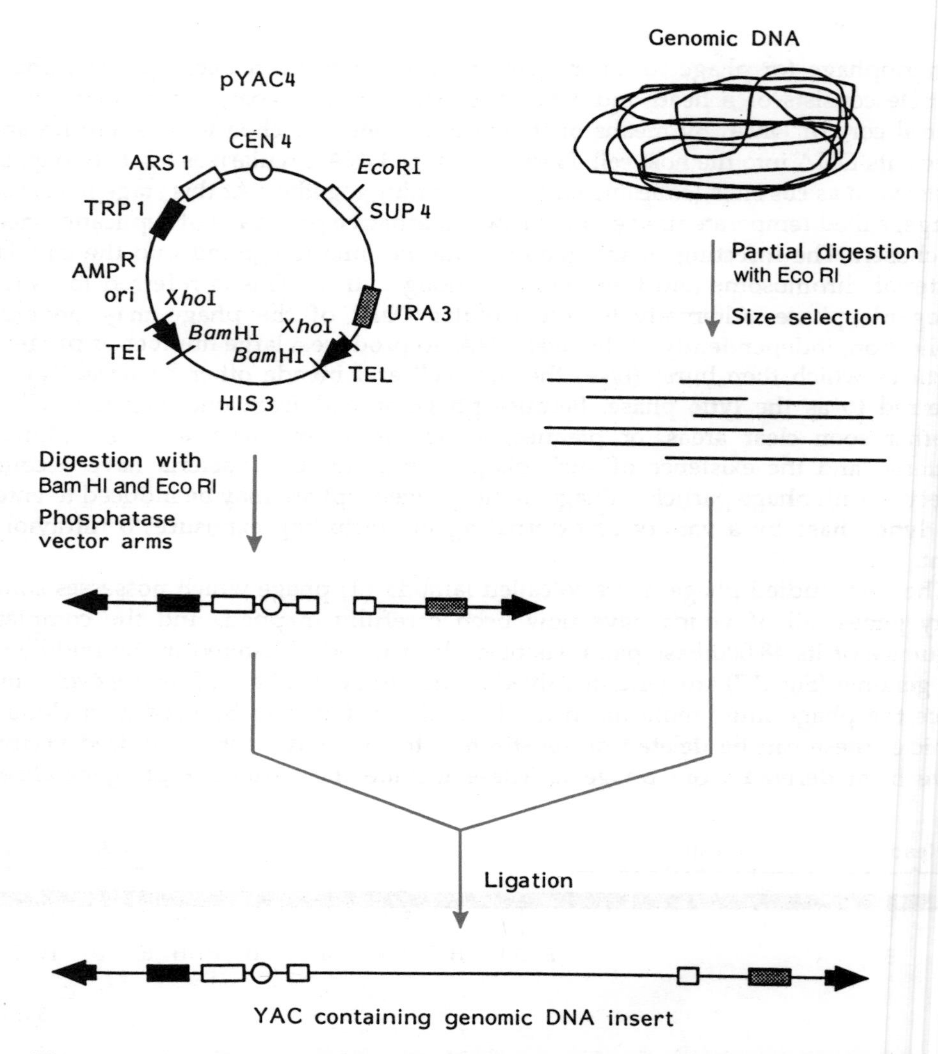

Figure 3.8 Construction of yeast artificial chromosomes. CEN, centromere; TEL, telomere; ARS, autonomous replicating segment; AMP^R, ampicillin resistance. (Courtesy Dr T. Woodage.)

Cosmids

A cosmid is an artificial construct of plasmid DNA packaged in a phage particle. They consist essentially of a circular duplex of DNA which contains various antibiotic resistance genes, several restriction sites, and *cos* sites of phage λ.

Like the selection of an appropriate restriction endonuclease, so the choice of a suitable vector depends on a number of factors. In general the technology is somewhat simpler using plasmids, the recombinants more stable and therefore easier to store in phage, and larger DNA fragments, up to 50 kb, can be cloned in cosmids.

Yeast artificial chromosomes

As gene mapping became increasingly popular and feasible there was, initially, a considerable gap between how close to the gene mapping studies could take one and the length of DNA piece which could be cloned in a conventional vector. Burke *et al.* (1987) realized that only a very limited amount of DNA was needed to provide all the functions necessary for a yeast chromosome to replicate stably. If telomere sequences, a centromere and an autonomous replicating segment (ARS) were present, the rest of the chromosome could be deleted and replaced by human DNA, hence the term artificial chromosome. A typical vector is shown in Fig. 3.8. It includes several other useful features such as a gene (TRP 1 or URA 3) to allow selection in medium lacking crucial factors and a gene to monitor for the presence of inserts by a colour change (SUP 4). DNA fragments of several hundred kilobases up to over a megabase have been cloned in YACs. Unfortunately such long pieces of DNA are often prone to rearrangements between different molecules resulting in chimeric YACs containing two separate DNA segments, but despite this disadvantage a large team of workers based at CEPH in Paris were able using YACs to construct the first complete map of the human genome in overlapping pieces (Cohen *et al.*, 1993).

CLONING

Having inserted a segment of foreign DNA into a suitable vector, the next step is to introduce the vector into a host organism. The host organism most often used is *E. coli*, the most favoured strain being K12 which has been cultured in the laboratory for over fifty years.

In order to generate recombinant plasmids these are first isolated from a bacterial host by disruption and centrifugation. The bacterial DNA and debris are discarded. The plasmid DNA and foreign DNA are then exposed to an appropriate restriction enzyme, mixed together and treated with ligase to produce a recombinant plasmid, sometimes also called a **chimaera**, named after the Greek mythological monster which, according to Homer, had the head of a lion, the body of a goat and the tail of a dragon. To reduce the likelihood of the plasmid recircularizing without incorporating the foreign DNA, the former is treated with the enzyme alkaline phosphatase which prevents this happening (Table 3.2). The recombinant plasmid

DNA is introduced into a bacterial cell (by treating the latter with calcium chloride which renders its cell membrane permeable to the plasmid DNA), a process referred to as **transformation**. The transformed *E. coli* is then grown on nutrient agar in Petri dishes. The process is illustrated diagrammatically in Fig. 3.9.

Under optimum conditions bacteria divide about once every 20 minutes so that after a day or so there will be literally millions of cells. At first these will appear as discrete colonies on the agar but if the culture continues the colonies will eventually fuse together to form a confluent layer of bacteria. The important point is that each colony of cells is the progeny of a single cell and therefore all the cells in a colony have the same genetic constitution. This is called a **clone** of cells and all the cells in a clone will contain the same inserted segment of foreign DNA. If such a colony is then picked off from the agar on which it is growing and transferred to another Petri dish and cultured, **all** the colonies in the subsequent culture will contain copies of the same DNA sequence. This process is called cloning and in this way it is possible to generate multiple (usually millions) copies of a particular recombinant DNA sequence.

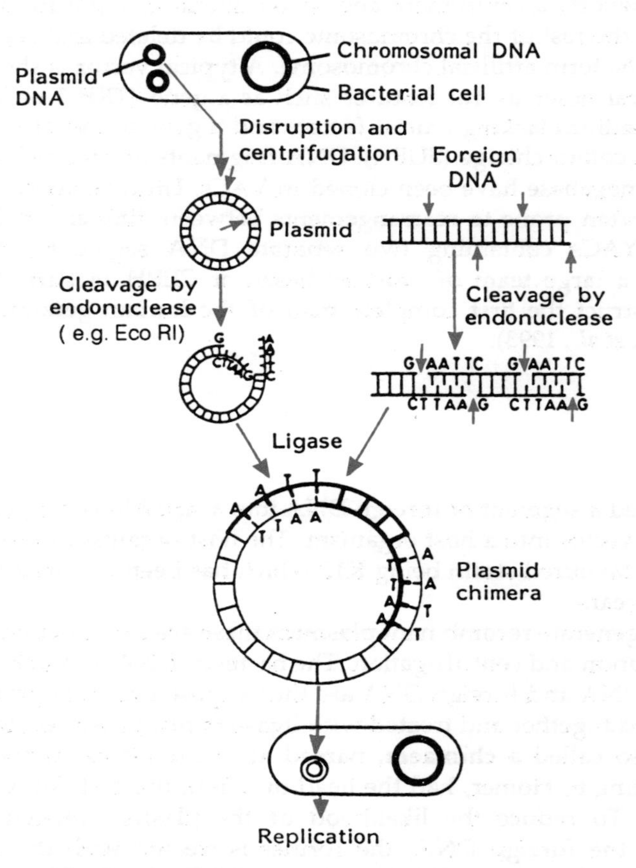

Figure 3.9 Generation of a recombinant plasmid.

SELECTION OF RECOMBINANT CLONES

The next problem is to determine if the attempt to produce recombinant clones has been successful and, if so, to select those clones which contain a specific DNA sequence. Some of the methods used for these purposes are summarized in Table 3.3.

The first step is to select clones which have incorporated the vector and have thus become transformed (transformation of a cell with DNA from a phage is sometimes called **transfection** to emphasize that the DNA is 'infectious'). Thus if the clone has become transformed by a phage this will be revealed by the formation of plaques, or if by a plasmid carrying resistance to a particular antibiotic by the ability of the clone to grow in the presence of the antibiotic.

The colonies are grown on a bed of agar, containing an antibiotic as appropriate, prepared in a Petri dish. The second step is to determine whether most of the vectors have acquired foreign DNA. The cloning site where the foreign DNA has been inserted is right in the middle of a gene, *lac*Z, which is required to code for β-galactosidase. The *lac*Z gene is immediately downstream from a promoter sequence and is under its control. If an inducer molecule, IPTG, is included in the culture medium this gene will be switched on. It will turn colonies blue by hydrolysing a chromogenic substrate, *Xgal*. However, if the enzyme has been inactivated by the successful insertion of cloned DNA the colonies will remain white.

The next step is to identify and characterize those recombinants which contain a specific DNA sequence. Here several approaches have been used but hybridization with an appropriate probe has proved by far the most useful. Replica copies of the library are made by laying a circle of nylon membrane over the dish and lifting colonies off, having marked the plate in some way so that the filter and plate can be orientated and reference eventually made back to the original master plate. The colonies are fixed on to the filter by baking and the DNA in each colony exposed by lysing with sodium hydroxide.

It is possible to identify clones containing a specific foreign DNA sequence by hybridization with an appropriate probe. A **probe** is a segment of single-stranded

Table 3.3 Summary of some of the methods used for identifying, selecting and characterizing clones

I. Selection of clones with a vector
   (a) Antibiotic resistance
   (b) Plaque formation

II. Selection of clones with a recombinant vector
   (a) Antibiotic resistance
   (b) *Xgal* test

III. Selection of recombinant clones with a specific DNA sequence
   (a) Genetic
      Complementation
      Immunological
   (b) Hybridization with an appropriate probe
   (c) Electron microscopy

DNA or RNA which has been labelled either radioactively or with some biochemical marker and is so named because it will search out and detect complementary sequences in the presence of a large amount of non-complementary DNA. The production of probes and their labelling will be dealt with later (page 48). The point here is that an appropriate probe will hybridize only with those colonies which possess complementary DNA. Colonies found to hybridize with a specific probe can be identified by autoradiography and then their location on the master plate determined. The method is shown in Fig. 3.10, and has been used for plasmid as well as phage vectors.

DNA blotting

Hybridization of DNA extracted from a clone can also be carried out on an electrophoresis gel using the method introduced by Southern and therefore referred to as a **Southern blot** (Southern, 1975). This is a very important procedure with wide applications in DNA technology. The method is shown in Fig. 3.11.

Extracted DNA is first exposed to an appropriate restriction endonuclease and the resultant fragments subjected to electrophoresis on an agarose gel. The gel is subsequently transferred from the electrophoresis tank to a dish containing sodium chloride/sodium hydroxide solution in order to denature the DNA. The gel is then rinsed and neutralized and placed on a filter paper wick in a bath containing SSC (sodium chloride/sodium citrate) buffer solution. On top of the gel is placed a nitrocellulose or nylon filter to which DNA binds strongly. Several layers of dry filter paper are then placed on top of the filter and held down by a heavy weight.

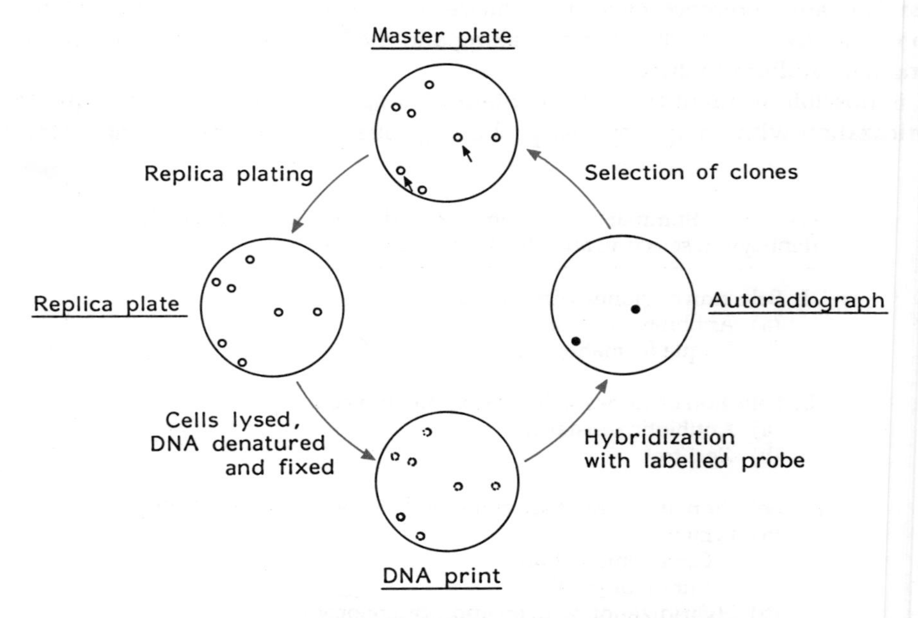

Figure 3.10 Detection of recombinant clones by hybridization with a radioactively labelled gene-specific probe.

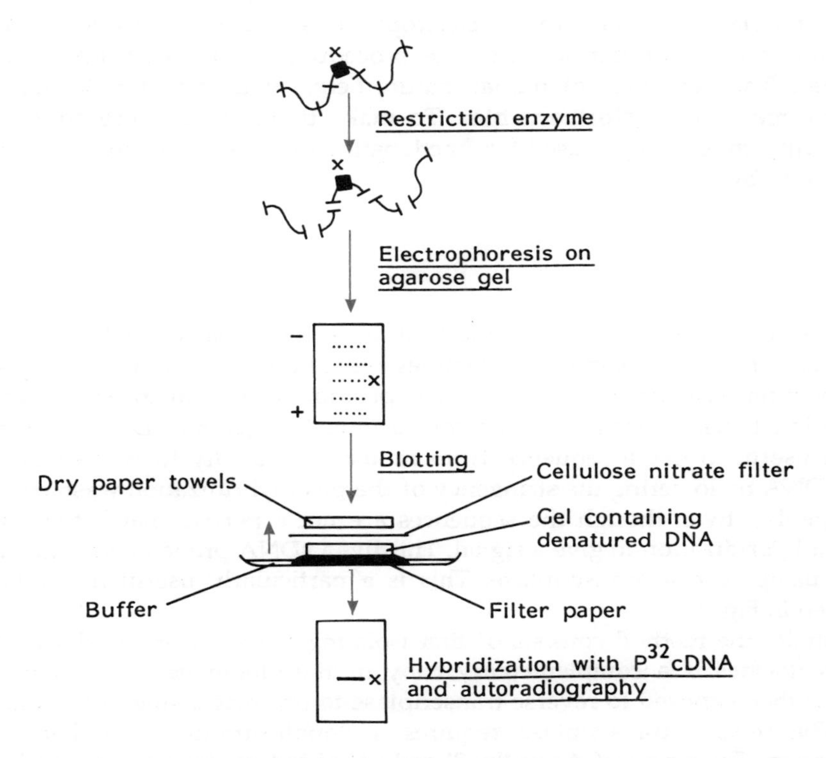

Figure 3.11 The Southern blot technique. x indicates the gene or sequence of interest.

The buffer is drawn by the dry filter paper through the gel, carrying with it the denatured single-stranded DNA fragments. When the DNA comes in contact with the filter it binds to it strongly. The DNA fragments are then permanently fixed to the filter by baking it in an oven at 80 °C in a vacuum. In this form the filter can be stored. Alternatively, it can be placed in a solution of an appropriate probe of labelled denatured DNA or RNA for hybridization to occur, after which the filter is washed several times in a buffer solution. The more stringent the washing conditions which are tolerated without removing any hybridized probe, the greater the degree of homology between the probe and the DNA bound to the filter. Hybridization with the radioactively labelled probe is detected by auto-radiography.

This simple description eschews the great care that is required if clear and reproducible results are to be obtained. The conditions of electrophoresis have to be very carefully controlled, the agarose gel has to be handled very gently because it is quite fragile, the filter must not be touched with bare hands because of the risk of contamination with extraneous DNA and if a radioactively labelled probe is used then there are the concomitant radiation hazards to be considered. Nevertheless the method has found much favour because of its relative simplicity and also because it is extremely sensitive and will detect sequences only a few hundred bases long among all the DNA in a complex genome. The method, however, cannot be applied to RNA fragments separated by gel electrophoresis because RNA does not bind to nitrocellulose and a modification of the procedure for use with RNA has been developed. The perversity of human nature being what it is, this technique has been referred to as a **Northern blot**. To make things even more confusing an entirely different procedure used for fractionating proteins is sometimes referred to as a **Western blot**!

GENE PROBES

A gene-specific probe may be produced in several ways. Firstly, it may be synthesized from its constituent nucleotides either because these are already known or can be inferred from the amino acid composition of the gene product. Secondly, it is possible to use an isolated and cloned segment of genomic DNA as a probe. It is often useful to use a sequence from mouse or fruit fly to cross-hybridize to human DNA by lowering the stringency of the post hybridization wash conditions (see page 47). Even though the sequences are not a perfect match there will be sufficient hybridization to give a signal. Thirdly, a cDNA probe may be made from mRNA using reverse transcriptase. This is a particularly useful method and is illustrated in Fig. 3.12.

Essentially the method consists of first isolating mRNA from a relevant tissue, say reticulocytes if a complementary copy of the globin gene is required. The mRNA is then exposed to reverse transcriptase to produce a single-stranded DNA copy. The reverse transcriptase requires a double-stranded template to start transcription. The string of As at the 3' end of mRNA (poly A tail) is exploited to provide a suitable starting point. A synthetic oligonucleotide made up of Ts is mixed with the messenger RNA. As this will base pair to the poly A tail transcription will start from there. This has the disadvantage that not all molecules

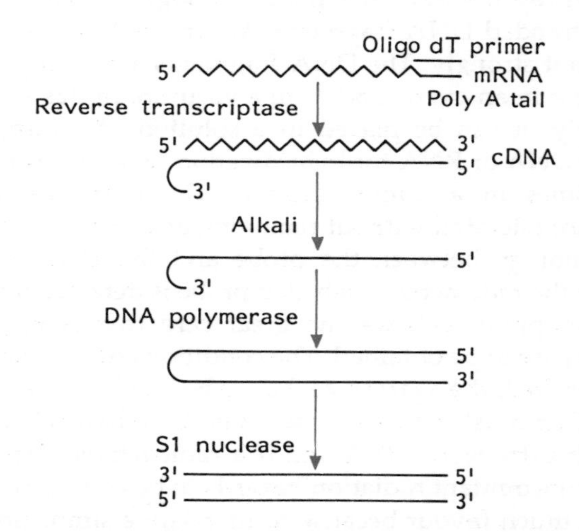

Figure 3.12 Synthesis of cDNA from mRNA using reverse transcriptase.

will be transcribed as far as their 5' end, particularly if the mRNA has some secondary structure. This is unfortunate as the coding information starts at the 5' end and in the majority of cases there is an untranslated 3' end which can be very long, up to half the molecule. Special measures such as use of internal primers and precautions to remove secondary structure are sometimes necessary. The RNA–DNA hybrid is then treated with alkali which destroys the RNA strand leaving the DNA strand intact.

In order to locate any DNA sequence with which a probe hybridizes, it is necessary to label the probe in some way. This may be achieved using either a radioactive marker (^{32}P) or a biochemical marker, and the label is incorporated into the DNA by a process referred to as **nick-translation**. Under appropriate conditions the enzyme DNase I produces small nicks at random along each DNA strand independently. The DNA molecule which has been nicked in this way is then exposed to DNA polymerase along with deoxynucleotide triphosphates labelled with, say, ^{32}P. The nicks form the foci for the DNA polymerase to act, the enzyme removing nucleotides from the 5' side of each nick and sequentially adding labelled nucleotides to the 3' side. In this way the reaction progressively incorporates a label into the DNA as the nicks are translated (=moved) along the molecule. Kits are now available for labelling DNA probes in this way.

An alternative method is to use a proteolytic fragment of DNA polymerase, the so-called Klenow fragment, which has lost its nuclease activity but is still active in polymerization. As explained for reverse transcriptase, a double-stranded template is required to prime transcription. This can be provided by heating the probe to make it single-stranded and then adding a mix of short oligonucleotides, six or nine bases long which will bind randomly over all the molecules. The Klenow fragment will then extend these molecules, faithfully copying the original sequence and by using a mixture of nucleotide triphosphates including radioactively labelled molecules a very 'hot' (i.e. high specific radioactivity) product can be obtained.

There is an understandable reluctance amongst many workers to use radioactivity, particularly ^{32}P, on a regular basis. There have been several efforts to replace it with non-isotopic methods. One method developed by David Ward and

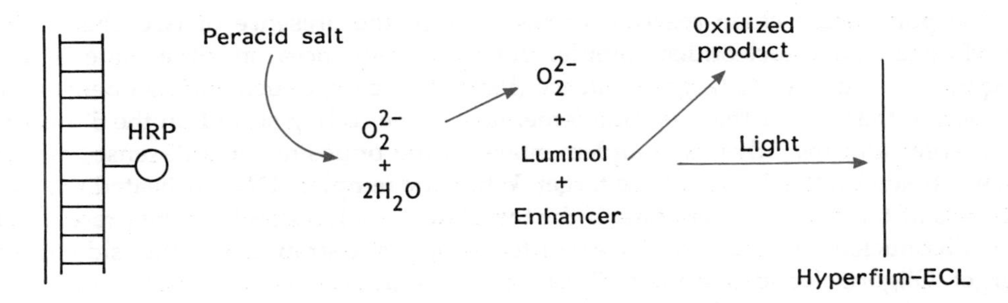

Figure 3.13 The principle of enhanced chemiluminescent detection (ECL). ECL systems are based on the generation of light, in the form of enhanced chemiluminescence to detect nucleic acids immobilized on membranes. The light production is catalyzed by horseradish peroxidase (HRP) bound to the nucleic acid probe. Results are achieved within one hour. Reproduced with permission from Amersham Life Sciences Catalogue, Amersham International p.l.c., Little Chalfont, Bucks, UK.

colleagues at Yale University uses a biochemical way of labelling the probe. Biotin, which has an extremely strong affinity for avidin, is covalently bound to the probe. The signal is detected by soaking the filter in avidin bound to a suitable enzyme and then using the enzyme in a colorimetric assay. Another recent innovation is to attach horseradish peroxidase to the probe (Fig. 3.13) and use this to produce a flash of light by chemiluminescence. The light is detected on X-ray film and the final result looks much like an autoradiograph. Although these methods may be safer than using radioactivity they are rather expensive, partly because of the large amounts of substrate needed to cover the whole filter, and they are generally not as sensitive.

THE POLYMERASE CHAIN REACTION (PCR)

If the title of this book is taken literally, the polymerase chain reaction has no place here as it avoids the need to clone DNA or make recombinant molecules. However, it has been so revolutionary in its applications and its uses follow so naturally from the methods described up to now that it forms an essential part of modern day recombinant DNA technology. It has rapidly acquired this core role because of its simplicity, sensitivity, speed, ability to tolerate low grade substrates, adaptability and non-radioactive applications.

The only technological advance needed for the invention of PCR was the ability to synthesize chemically short oligonucleotides containing up to 30 bases. The chemistry to do this gradually improved during the 1980s. After that it only required an ingenious application of standard nucleic acid biochemistry and enzymology by Kary Mullis to devise this method, which amplifies many thousand fold a specific section of DNA from amongst many sequences present. The problem lies in the complexity of DNA. For example, although each human cell contains 3 billion base pairs of DNA, the most common mutation causing cystic fibrosis is caused by the deletion of just three base pairs. As explained previously, that particular section can be isolated and amplified by cloning but this is time consuming and expensive.

The polymerase chain reaction works through the presence of two chemically synthesized oligonucleotides complementary to sequences on either side of the sequence of interest, or target sequence (Plate I). The oligonucleotide sequences are chosen so that one of them is complementary to the coding strand on the 5' side of the target and the other is complementary to the opposite (or anti-sense) strand downstream on the 3' side of the target. When the template DNA is heated the two strands of the helix will separate. If the template is cooled slowly in the presence of the oligonucleotides they will base pair to the opposite strands on either side of the target. Oligonucleotides around 20 base pairs are usually used as their sequence is sufficiently long that it should not appear randomly anywhere else in the genome. As explained before, this short double-stranded template can be extended by DNA polymerase in the presence of the four nucleotides to make a copy of the original sequence. As nucleotides are added at the 3' end of the molecule the two chains will grow in opposite directions and the result will be two complete double-stranded copies of the target.

The beauty of the PCR reaction is that the whole procedure can be repeated. If the temperature is raised again the two strands of the new molecules will dissociate, on cooling to the annealing temperature the oligonucleotides will again bind around the target, but this time they will bind to the original and to the newly synthesized molecules. The temperature can be cycled many times (giving rise to the alternative name thermal cycling) and there will be an exponential build up of molecules until one of the components is exhausted.

During the first cycle there is no theoretical limit to the length of the newly synthesized molecule. However, in subsequent rounds of amplification many of the template molecules will themselves have started with one of the oligonucleotides, so after the first few rounds of amplification most molecules will be of identical length. Practical examples will be found in later chapters. Originally the whole procedure was rather cumbersome with addition of fresh enzyme during each cycle. However, the introduction of a thermal stable polymerase from *Thermophilis aquaticus* (or *Taq*I) simplified the procedure enormously and made automation possible. In theory a single molecule is sufficient to start the whole process off. Although this is technically demanding, it is possible and has been used for preimplantation diagnosis. In general degraded DNA, perhaps from a paraffin block stored in a pathology department for many years, will have a few molecules sufficiently long to stretch between the two sites for oligonucleotide binding and intact product can be rescued. Some examples of sources of DNA which have been used as templates in the PCR are shown in Table 3.4. One method which helps to overcome the problems associated with very small amounts of target DNA is the use of nested primers. PCR relies critically on the accurate binding of oligonucleotides around the target and sometimes, despite careful design of primers, only a partial match may form. Any such non-specific molecules will be amplified through subsequent cycles just the same as the correct molecule. If there are very few starting molecules this could produce a major problem. Therefore, a second set of primers are designed within the amplified sequence (hence the term nested). After initial PCR cycles some of the product is taken and reamplified with the second set. Only the correct molecules will have binding sites for this second set of primers and true replication should be restored (Fig. 3.14).

A most useful application of PCR is to combine it with reverse transcription of mRNA. RNA is more difficult to work with than DNA, not only because of its

Table 3.4 Sources of human DNA used in polymerase chain reaction

White cells from blood
Buccal smear
Mouth wash
Semen
Paraffin blocks
Bone from archaeological remains
Cervical smears (for presence of papilloma virus)
Tumour cells

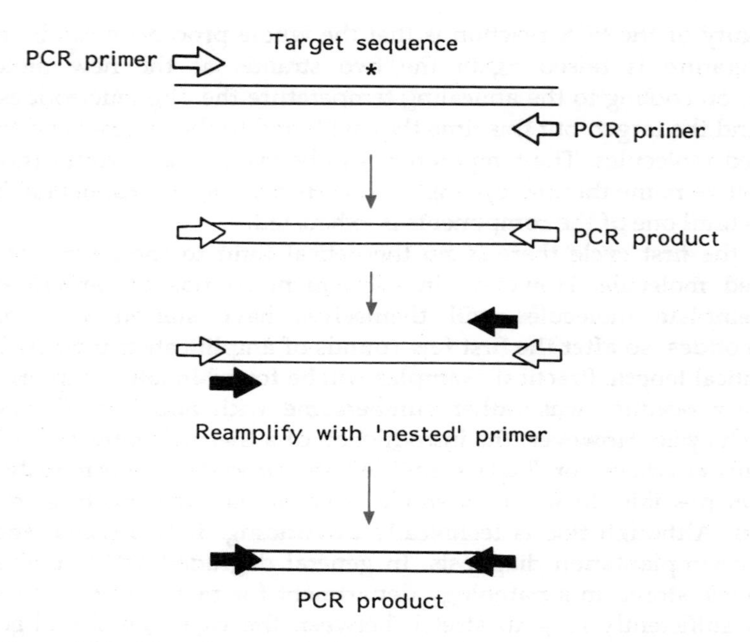

Figure 3.14 Nested primers in PCR.

increased lability but because the correct tissue must be studied in which the mRNA is expressed. If mRNA is reversibly transcribed into cDNA as described previously (p. 48) then the resulting single-stranded cDNA molecule is a suitable starting template for PCR. mRNA species of very low abundance can be amplified using this method which is called RT–PCR. A PCR product made in this way can be used as a probe in subsequent experiments.

PCR has, of course, many applications in the study of infectious disease as well as genetic disease because of its ability to detect very small amounts of sequence. A fascinating use recently involved the analysis of a mummy from the pre-Incan Chiribaya culture of southern Peru. In the very dry climate of that area the body had dried out to form a natural mummy. PCR was used to isolate specific sequences of *Mycobacterium tuberculosis* from the body which shows that tuberculosis was present in the New World before Europeans arrived.

MONOCLONAL ANTIBODIES

The production and use of monoclonal antibodies do not involve recombinant DNA technology but since the two technologies can complement each other it is perhaps convenient at this point to mention the subject. The technique of producing monoclonal antibodies was pioneered by Milstein of the Laboratory of Molecular Biology at Cambridge. Though various methods differ in detail essentially the technique consists of taking myeloma cells, which are capable of growing indefinitely in tissue culture, and fusing these with spleen cells (B lymphocytes) obtained from a sensitized animal. Each one of the lymphocytes is capable of making a particular antibody but will only grow in culture when fused with myeloma cells. From such hybrid cells (or **hybridomas**) separate clones can be isolated, each of which will constitute the descendants of one lymphocyte and therefore will synthesize one particular antibody. The isolated clones can then be cultured further and in this way significant amounts of a particular antibody can be produced. This technology has many uses. It is possible, for example, to study individual antigens which are jumbled together on a cell surface. It is therefore being used to identify surface antigens on cancer cells which might then be employed for diagnosis, and perhaps the production of appropriate monoclonal antibodies from such cells could be used in treatment. Anti-T-cell monoclonal antibodies may prove useful for preventing graft rejection.

The interest from the present point of view is that a labelled monoclonal antibody may be used in conjunction with DNA technology to identify, for example, the location and distribution of particular peptides in histological sections of the central nervous system using *in situ* hybridization methods.

SUMMARY

The recombinant part of recombinant DNA technology can be considered under four main headings.

- Firstly, the generation of DNA fragments using sequence-specific restriction endonucleases.
- Secondly, the incorporation of these fragments into a suitable vector such as a plasmid, phage or cosmid.
- Thirdly, the introduction of the vector into a particular host organism, usually *E. coli*, which is then grown in culture to produce clones with multiple copies of the incorporated DNA fragment.
- Finally, the identification, selection and characterization of recombinant clones for which a number of ingenious techniques have been devised, including hybridization on a Southern blot using a labelled gene-specific probe.
- The polymerase chain reaction often replaces the need to clone a molecule because of its ability to produce large amounts of specific DNA sequences rapidly.

REFERENCES AND FURTHER READING

Textbooks and review articles

Emery AEH. *Elements of Medical Genetics*, 6th edn. Edinburgh: Churchill Livingstone, 1983

Emery AEH, Mueller R. *Elements of Medical Genetics*, 8th edn. Edinburgh: Churchill Livingstone, 1992

Maniatis T, Fritsch EF, Sambrook J. *Molecular Cloning—A Laboratory Manual*. Cold Spring Harbour Lab., 1982

Rosenthal N. DNA and the genetic code. *New Engl J Med* 1994; **331**: 39–41.
Rosenthal N. Tools of the trade—recombinant DNA. *New Engl J Med* 1994; **331**: 315–317.
Rosenthal N. Stalking the gene—DNA libraries. *New Engl J Med* 1994; **331**: 599–600.
Many volumes containing useful practical details are available in the series *Practical Approaches to* ... IRL Press Oxford

Research publications

Benton WD, David RW. Screening λgt recombinant clones by hybridization to single plaques *in situ*. *Science* 1977; **196**: 180–182
Bolivar F, Rodriguez RL, Greene PJ, Betlach MC, Heyneker HL, Boyer HW. Construction and characterization of new cloning vehicles II. A multipurpose cloning system. *Gene* 1977; **2**: 95–113
Burke DT, Carle GF, Olson MV. Cloning of large segments of exogenous DNA into yeast by means of artificial chromosome vectors. *Science* 1987; **236**: 806–812
Cohen D, Chumakov I, Weissenbach J. A first generation physical map of the human genome. *Nature* 1993; **366**: 698–701
Grunstein M, Hogness DS. Colony hybridization: a method for the isolation of cloned DNAs that contain a specific gene. *Proc Natl Acad Sci USA* 1975; **72**: 3961–3965
Leary JJ, Brigati DJ, Ward DC. Rapid and sensitive colorimetric method for visualizing biotin-labelled DNA probes hybridized to DNA or RNA immobilized on nitrocellulose: bio-plots. *Proc Natl Acad Sci USA* 1983, **80**: 4045–4049
Southern EM. Detection of specific sequences among DNA fragments separated by gel electrophoresis. *J Mol Biol* 1975; **98**: 503–517

Chapter 4
Gene mapping, structure and function

Our views of the structure and function of genes have had to undergo major changes in the last few years as a result of findings with recombinant DNA technology. Genes can no longer be considered simply as discrete and contiguous units on a chromosome. The picture is very much more complicated, and what has been learned is beginning to throw considerably more light on our understanding of the cause and pathogenesis of human disease. But first it is necessary to consider in some detail the arrangement and structure of normal genes.

GENE MAPPING

Human somatic cells have 46 chromosomes composed of one pair of sex chromosomes (XX in the female, XY in the male) and 22 pairs of autosomes. The two members of any given pair of autosomes are said to be **homologous**, one homologue of a pair being derived from one parent and its partner from the other parent.

By convention the short arm of a chromosome is represented by 'p' and the long arm by 'q'. By special staining with, for example, Giemsa, each chromosome exhibits a specific banding pattern. Based on these banding patterns each arm of a chromosome is divided into regions and each region into bands, numbering always from the centromere outwards (Fig. 4.1). A given point on a chromosome is designated by the chromosome number, the arm symbol, the region number and band number in this order (*Cytogenetics and Cell Genetics* 1978; **21**: 313–404). Thus the gene for Duchenne muscular dystrophy is located at Xp21 (Fig. 4.2).

When different genes are located on the same chromosome pair they are said to be **linked**. As a result of the phenomenon of crossing-over during meiosis in gamete formation, there is an exchange of material between homologous chromosomes, and the relative distance between genes on any particular chromosome is measured by the frequency with which crossing-over occurs between them. Distances between genes are measured in **map units**, one map unit being equal to one per cent crossing over. A map unit is frequently referred to as a **centiMorgan** (cM), so named after Thomas Hunt Morgan who, in the 1930s, pioneered work on gene mapping in the fruit fly *Drosophila*. A centiMorgan is roughly equal to a thousand kb or 10^6 base pairs.

A long-standing method of gene mapping, so-called linkage mapping, involved studying families to determine the frequency with which crossing-over occurred

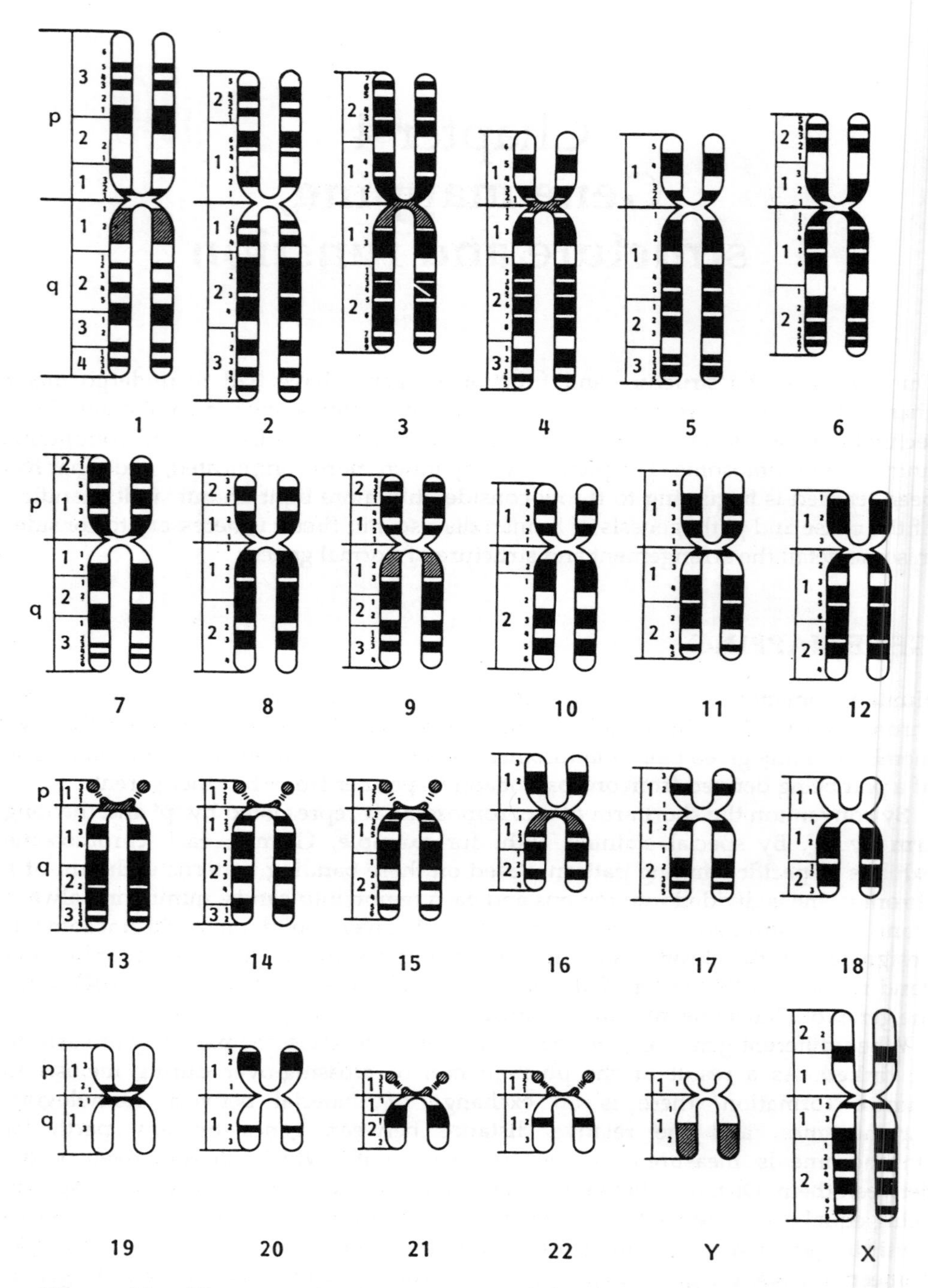

Figure 4.1a Diagrammatic representation of banding patterns of individual chromosomes revealed by fluorescent and Giemsa staining. (From Davidson, 1978, with permission.)

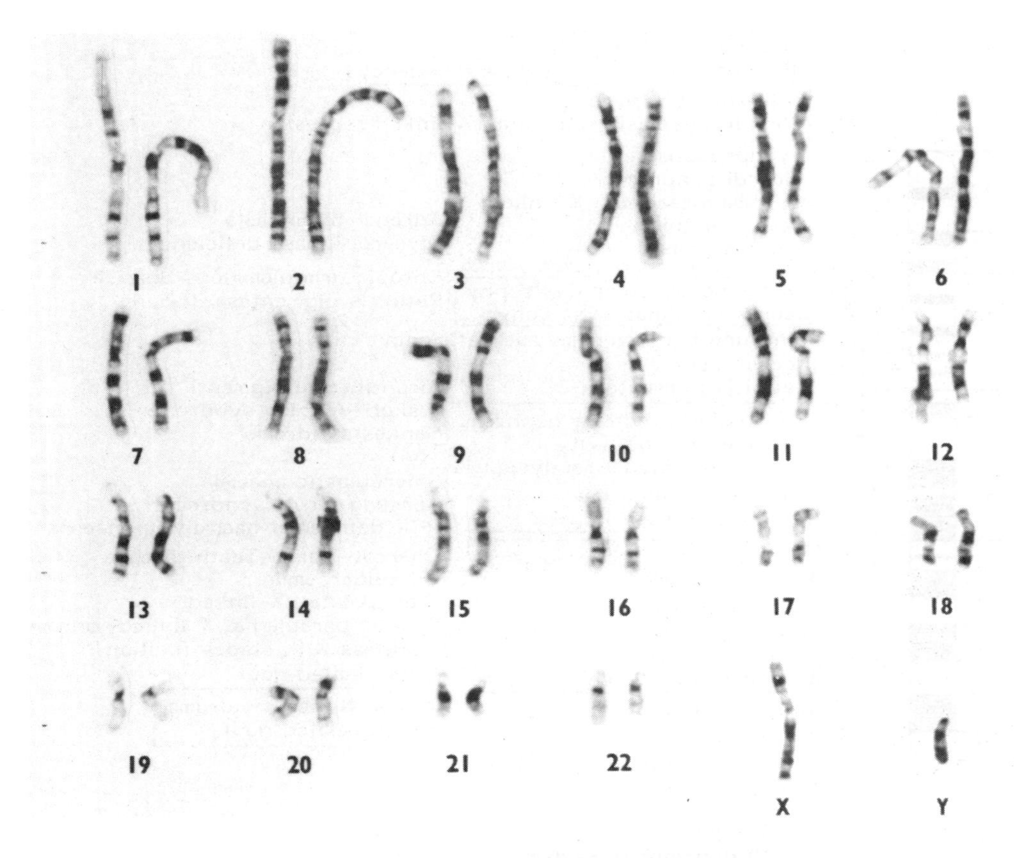

Figure 4.1b Photograph of human chromosomes.

between a gene for a particular disease and genes for various genetic markers, traditionally blood groups and certain serum proteins.

The general principle of linkage analysis is shown in Fig. 4.3. This is considerably easier in mice and other fast breeding animals than in humans as real human families rarely have the pedigree structure desirable for powerful statistical analysis. A number of computer programs have been developed to analyse the chance of linkage (i.e. the chance that two traits are genuinely co-inherited more often than at random) in human pedigrees. As multiple DNA based genetic markers, based on short tandem repeats, were developed for all human chromosomes. Linkage analysis became very powerful and many diseases, e.g. cystic fibrosis, neurofibromatosis and colonic cancer, were first mapped this way. Although the human genome is now richly covered with highly informative PCR formatted markers the number of single gene disorders for which adequate sized families can be collected may be nearly exhausted. Attempts to map disorders such as schizophrenia in which the pattern of inheritance is not clear and the diagnostic criteria controversial have had mixed success.

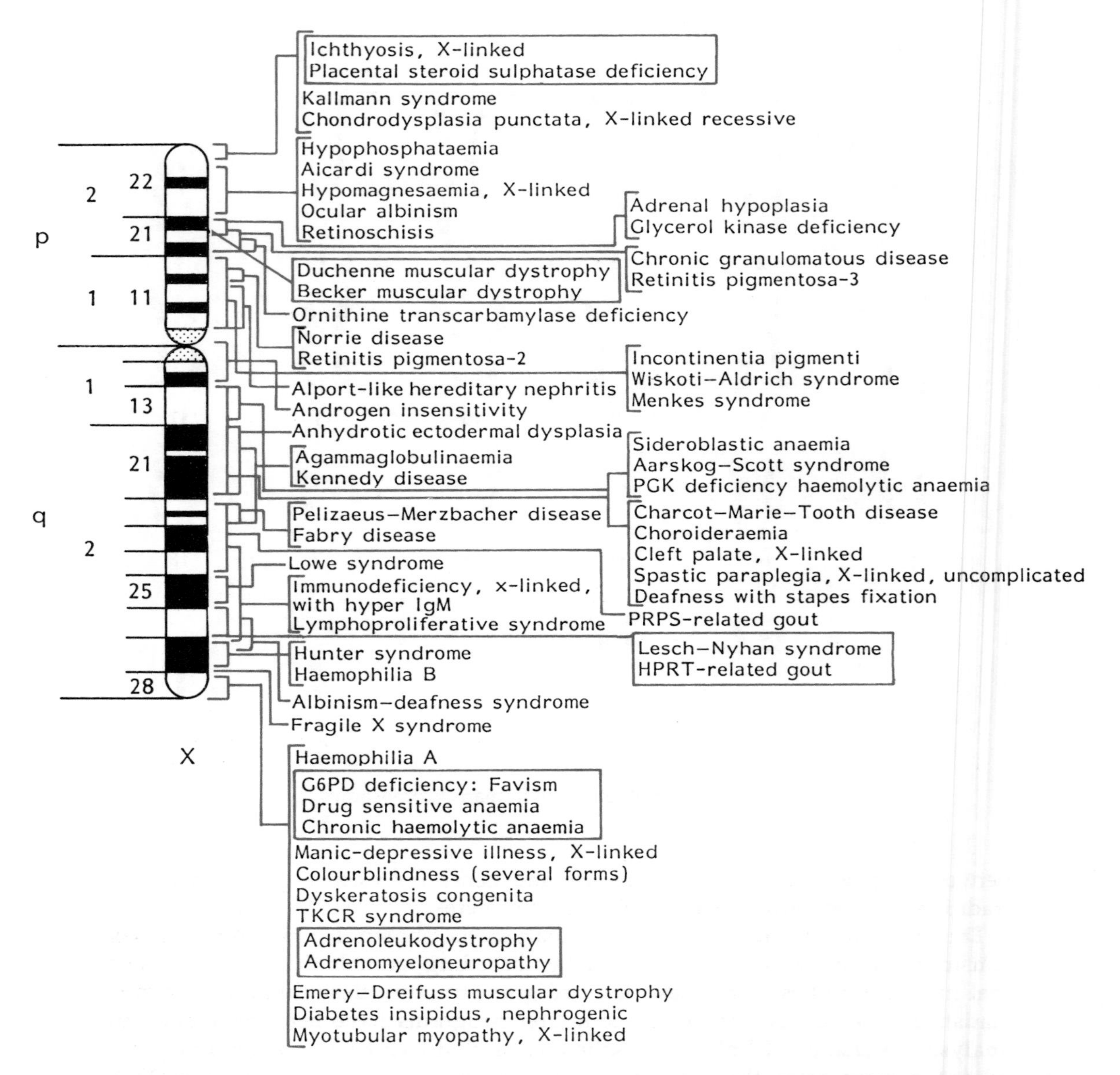

Figure 4.2 Simplified map of the human X chromosome. (From McKusick, 1992, with permission.)

Tracking of genes through families using known closely linked markers is of great importance in molecular diagnostic laboratories and the subject will be covered in greater detail in Chapter 7.

Somatic cell genetics has also provided useful information. In this latter approach cells from a human and rodent are fused together (using a virus or polyethylene glycol) to produce **somatic cell hybrids**. As such hybrid cells divide the human

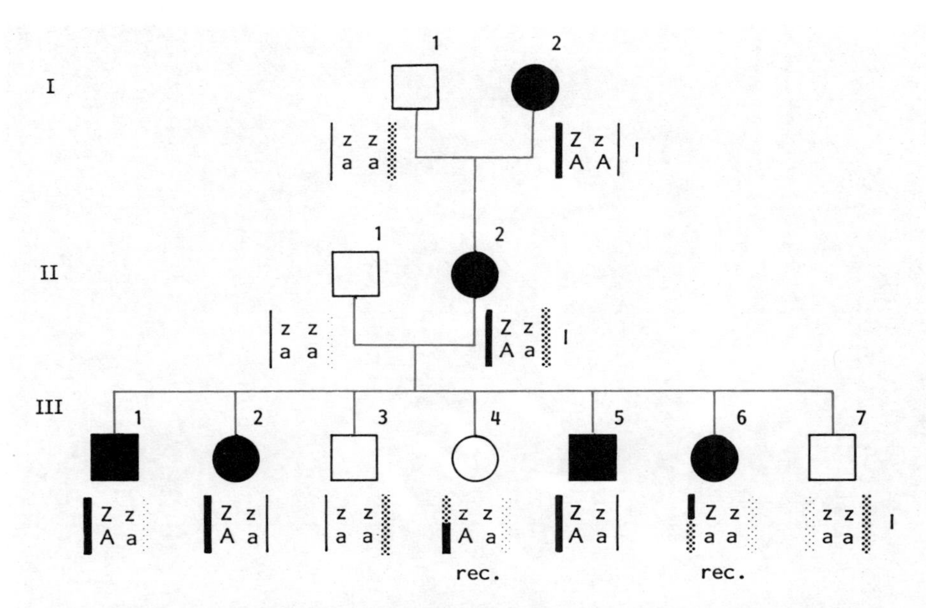

Figure 4.3 Principles of linkage analysis. It is possible to trace the analysis of the genetic markers Z/z and A/a and the disorder through the family from the affected grandmother to the grandchildren. In individuals III-1, 2 and 5 the intact chromosome with alleles ZA have been passed on, together with the disease. In individuals III-3 and III-7 the chromosome originating in the normal grandfather, za, has been passed on with no disease. However, in individual III-6 there has been a recombination during meiosis in her mother, II-2, and she has received a recombinant chromosome, Za, together with the disease. The disease allele is travelling with Z but not a. Individual III-4 has also inherited a chromosome in which there has been recombination between z and a. She has inherited zA and no disease. The disease is again travelling with the Z marker in this and every other individual, thus showing linkage.

chromosomes are gradually lost and correlations can be made between the presence of human DNA sequences or genes and the remaining chromosomes in the cell. Early experiments used various biochemical markers exhibited by the cells in tissue culture but this approach was limited by the cell specific expression of many genes. Labelled DNA probes were much used to probe the DNA from hybrid cells using the Southern blotting technique. By studying the hybridization patterns of different hybrid cell lines possessing different complements of human chromosomes, it has been possible to assign different genes to specific chromosomes. If the initial hybrids are made from cells with chromosomal translocations the gene position can be pinpointed to a specific arm or chromosomal region. Blotting methods have largely been replaced by PCR technology. A PCR product will only be obtained if the chromosome carrying the gene under study is still present.

A particularly attractive method of localizing a gene is by *in situ* hybridization. Here a chromosome spread fixed to a microscope slide is exposed to a labelled gene probe. If the DNA strands of the fixed chromosome have been separated by heating, the probe will hybridize to complementary sequences in the chromosomes

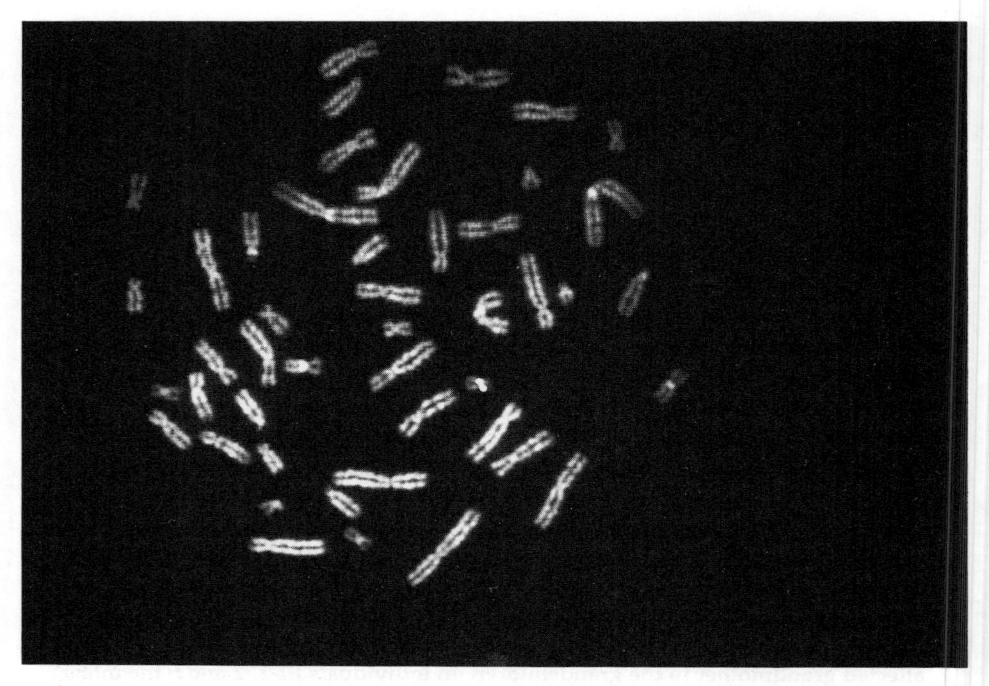

Figure 4.4 Fluorescent *in situ* hybridisation (FISH). A probe from chromosome 22 has been used to diagnose a case of di George syndrome, in which one of the copies of chromosome 22 carries a deletion. Only one copy of chromosome 22 (centre) hybridises to the probe.

and their location can be detected by microscopy. If the DNA probe is biotinylated, avidin bound fluorescent molecules can be used to detect the signal. This is called fluorescent *in situ* hybridization or FISH (Fig. 4.4).

POSITIONAL CLONING

The underlying impetus for mapping a disease-causing gene is to provide a starting point for the isolation of the gene (Collins, 1992). This is particularly important if the gene product causing the disease is a previously unknown gene. Unfortunately, as can be seen in Fig. 4.5 the resolution of the mapping methods just described usually leaves a large gap (several orders of magnitude) between gene localization and most cloned molecules. Yeast artificial chromosomes described in Chapter 3 help to close the gap. A typical approach used in positional cloning (Fig. 4.6) would be to use a pair of PCR primers, which have been used to detect a closely linked polymorphic marker in a family linkage study, to isolate a YAC which contains the sequence. This is done by pooling sets of YACs and finding a pool which gives a positive PCR product. The original pool is then subdivided and so on. The primer sequences are called sequence tagged sites or STSs and are useful map coordinates. Once a YAC has been identified the corresponding cosmids are isolated. Whereas each YAC contains several hundred kilobases of DNA and is cumbersome to

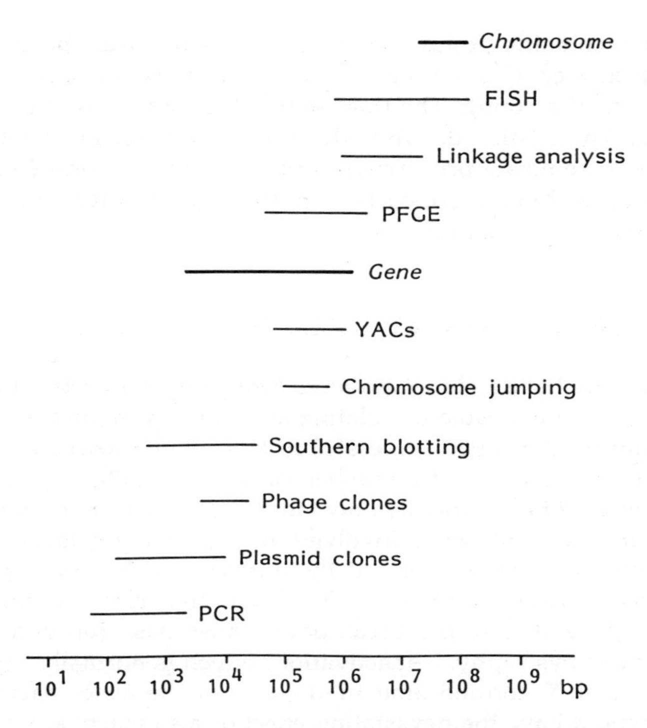

Figure 4.5 Level of resolution of different cloning and mapping methods. Diagrammatic representation of the approximate effective range of a number of commonly used DNA cloning and mapping methods. The size ranges of typical mammalian genes and chromosomes are shown for comparative purposes. (Courtesy of Dr T. Woodage.)

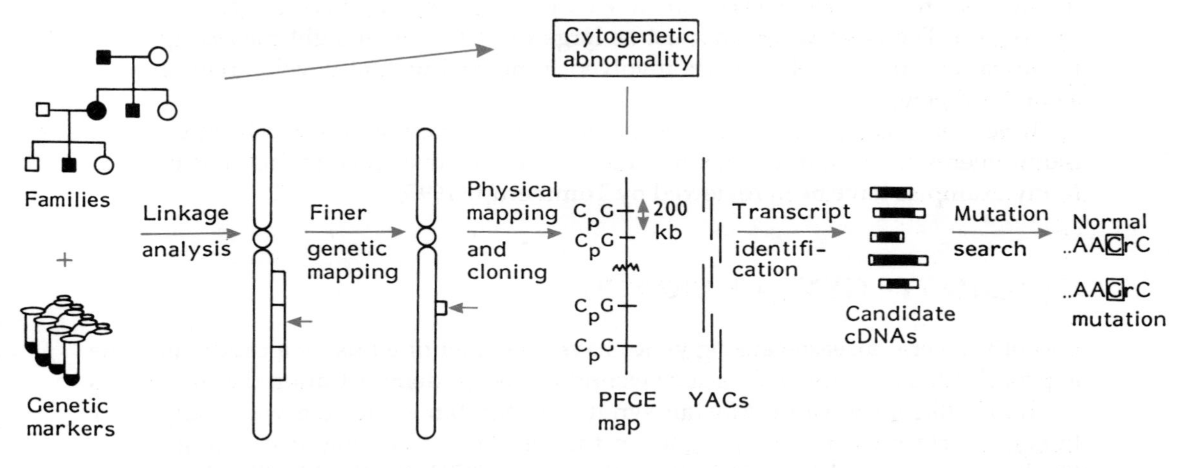

Figure 4.6 Diagram of the approach generally utilized for positional cloning. (From Collins, 1992, with permission.)

handle a cosmid contains around 40 kb of DNA and can be grown in large amounts. The next step is to order the set of cosmids into an overlapping, or contiguous set, called a contig. The final step is to identify coding regions within the cosmid DNA. This is still difficult to do and methods for improving this part of gene identification are now a priority within the Human Genome Project. Methods currently available will be described later in the chapter after more properties of human genes have been explained.

CHROMOSOME REARRANGEMENTS

Rare individuals who have a chromosome rearrangement associated with a genetic disorder have proved invaluable in isolating and identifying disease-causing genes ever since this approach proved successful in Duchenne muscular dystrophy. The first such examples were a small number of women suffering from Duchenne muscular dystrophy, which normally only affects boys, who were noticed to have a chromosome rearrangement which involved an apparently balanced translocation between the X chromosome and one of the autosomes. The break-point on the X chromosome was always within band Xp21 but the other chromosome varied widely. It was realized that the break-point must pass through the gene for Duchenne muscular dystrophy. X-inactivation, which is normally random, mainly involves the normal X chromosome in these cases because inactivation of the translocated X would have the devastating effect of also inactivating the reciprocal autosomal chromosome. Thus the normal Duchenne (or dystrophin) gene is inactivated by X-inactivation and the other is inactivated by disruption by translocation. The women end up with no active dystrophin genes.

A further example of a useful rearrangement was found during the hunt for the gene for familial adenomatous polyposis (FAP). A child with both mental retardation and a sporadic case of colon cancer was observed to have a small deletion of the long arm of chromosome 5. Linkage studies rapidly followed using markers from the deleted area and showed that the gene in affected families lies in this region. The deleted region must be large to be visible by light microscopy and presumably removes other genes involved in normal mental development as well as the FAP gene.

These translocations are very useful to gene hunters as obviously one of the requirements for a candidate gene is that it should be disrupted in the translocation. Many examples have been reviewed by Tommerup (1993).

CANDIDATE GENE APPROACH

Recently several disease-causing genes have been identified using a candidate gene approach (Ballabio, 1993). One such example was the demonstration that mutations of the fibrillin gene cause Marfan syndrome. Marfan syndrome was mapped by linkage to chromosome 15, roughly in the middle of the long arm. The gene for fibrillin was mapped to a similar position using FISH. As the fibrillin structure is disordered in the skin and fibroblasts of patients with Marfan syndrome the gene was a strong candidate. DNA sequencing of the fibrillin gene in a number of patients found mutations which would lead to inactive fibrillin.

A more unexpected example was the finding that a gap junction protein, connexin 32, causes the X-linked form of the peripheral neuropathy Charcot–Marie–Tooth (CMT) disease. The gene for connexin 32 was shown to map to exactly the region of the X chromosome found, by linkage studies, to contain the CMT gene (Xq13). DNA sequencing rapidly established it was the true gene. This was a surprising result for several reasons, one being that connexin 32 is expressed most strongly in the liver and there is no liver involvement in this peripheral neuropathy. This finding shows that the large investment in the Human Genome Project is already paying dividends in medical science.

THE TRANSCRIPTION UNIT

In eukaryotes almost all genes are interrupted by intervening sequences or **introns**, the remaining parts of the gene separated by introns being called **exons**. These terms were first introduced by Gilbert, one of the pioneers of DNA sequencing. It is only the exons which are transcribed into mRNA and therefore specify the gene product.

Evidence that genes in higher organisms are 'split' and rarely continuous stretches of DNA came from a number of sources. The first studies to show this in a mammal were based on restriction mapping. These investigators hybridized restriction fragments of rabbit liver DNA with a β-globin labelled probe. The restriction fragments so identified as being around and within the β-globin gene then enabled the investigators to draw up a physical map of the cleavage sites in the region. This revealed that the β-globin gene is not contiguous but is interrupted by a DNA segment inserted within the coding sequence. Otherwise the restriction map within the gene was colinear with that deduced from the known sequence of rabbit β-globin mRNA. That genes are interrupted in this way is also shown by comparing the base sequences of genomic DNA with mRNA (or cDNA). In this way it became clear that there are sequences in genomic DNA which are absent from mRNA. This incongruity between genomic DNA and mRNA can also be demonstrated visually, by electron microscopy. An RNA can be hybridized with a single-stranded DNA. Where the two molecules are complementary, a double-stranded hybrid will be formed which will appear thicker on electron microscopy than regions which are not complementary and therefore do not hybridize and so remain single-stranded. This technique can be used not only to study homologies between DNA and RNA but also between different DNAs. The method is sometimes called **heteroduplex mapping**. When RNA is hybridized with single-stranded DNA, regions of the DNA which do not hybridize are extruded as single-stranded loops. These extruded portions represent introns and from studying electron micrographs of such structures it was possible to determine their number, size and location within a transcription unit (Fig. 4.7).

Thus from experiments such as these it became clear that there are sequences within genomic DNA that are not represented in mRNA. A particularly ingenious method for demonstrating the presence of introns (as well as the ends of RNA

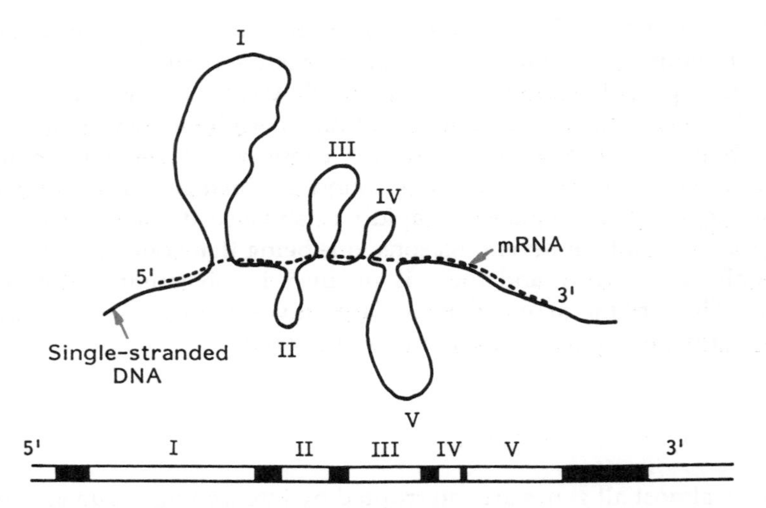

Figure 4.7 Diagrammatic representation of hybridization between single-stranded genomic DNA and mRNA as seen on electron microscopy. Below, the inferred arrangement of exons and introns in the transcription unit.

molecules) is variously referred to as **S1 nuclease mapping**, or Berk–Sharp mapping after its originators (Berk and Sharp, 1977). The method is illustrated in Fig. 4.8.

Genomic DNA is allowed to hybridize with mRNA and then treated with alkali to destroy the RNA. Electrophoresis of the resultant DNA fragments will provide the total size of the transcription unit including both introns and exons. However, if the hybrid is exposed to S1 nuclease this will remove the extruded single-stranded DNA, corresponding to the introns, and the resultant DNA–RNA hybrid will have a **small** gap in the DNA strand where the extruded DNA was excised. Subsequent treatment with alkali releases the fragments of DNA, which correspond to the individual exons, and their size (length) can then be determined on electrophoresis. If the hybrid is not treated with alkali after exposure to S1 nuclease, then the exons will be held together by the RNA and on electrophoresis the resultant fragments will reflect the combined size of all the exons in the gene.

The picture which emerged from these studies was that almost all genes in eukaryotes are split by introns, although in man there are just a few exceptions such as the gene for histones and the sex-determining gene, SRY. In other genes the number and size of the introns vary considerably. For example, each of the several collagen genes possess some fifty introns varying in size from 100 to 2500 base pairs. The globin genes are comparatively simple and each of the constituents in the α-globin gene region on chromosome 16 and the β-globin gene region on chromosome 11 had two introns and three exons (page 71). An exon can be of any length and may or may not be an exact multiple of three base pairs (i.e. one coding unit). The relevance of this will be seen in Chapter 7 discussing the difference between the Becker and Duchenne forms of muscular dystrophy.

The function of introns is not clear though it is suspected that they may be involved in some way with gene regulation. What is known, however, is that

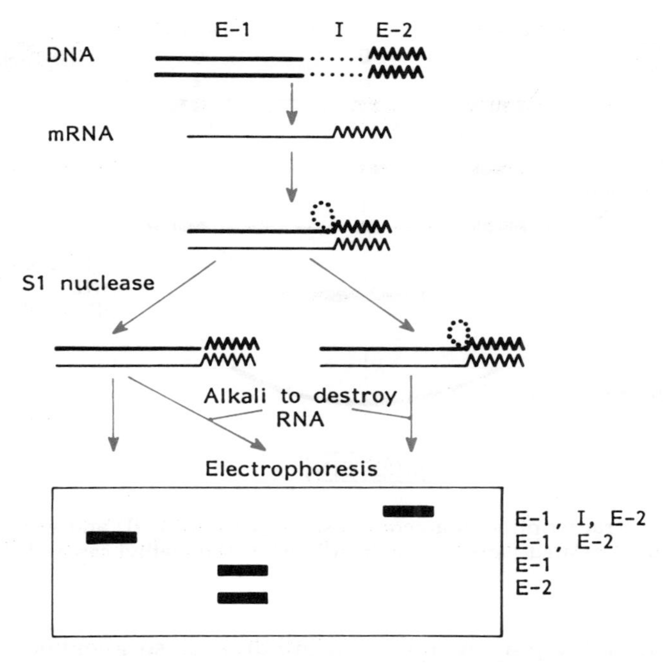

Figure 4.8 SI mapping for determining the lengths of exons (E-1, E-2) and introns (I) within a gene.

mutations within introns may seriously affect RNA splicing and thereby interfere with correct gene synthesis. There are many examples of such mutations having important clinical consequences.

The first step in the synthesis of mRNA is the transcription of the entire unit of both introns and exons into precursor RNA. The region of the precursor RNA transcribed from the introns is then excised and does not form mRNA and therefore does not specify the primary structure of the gene product. On the other hand, the precursor RNA from exons is not excised and is spliced together to form the definitive mRNA which specifies the primary structure of the resultant gene product (Fig. 4.9). It should be noted that the first two bases of the 5' end of each intron are invariably guanine and thymine and the last two bases are adenine and guanine, and these are essential for accurate excision and splicing.

Interestingly, when the first G in the intron is mutated, the usual consequence is that the whole preceding exon is not recognized and splicing occurs between the surrounding exons (A and C in Fig. 4.9) giving rise to exon skipping. The exact signals controlling precursor mRNA splicing are subtle but we know that the macromolecular complex controlling the whole process, the spliceosome, contains five small nuclear RNAs (snRNAs) and a whole array of proteins. The snRNAs base-pair with short regions of the introns, thus bringing the two exons physically close together.

Before intron transcriptions are removed and splicing occurs, a so-called **cap** of methylated guanosine residues is added to the 5' end of the precursor RNA (not to be confused with the CAP factor in prokaryotes which is quite different) and

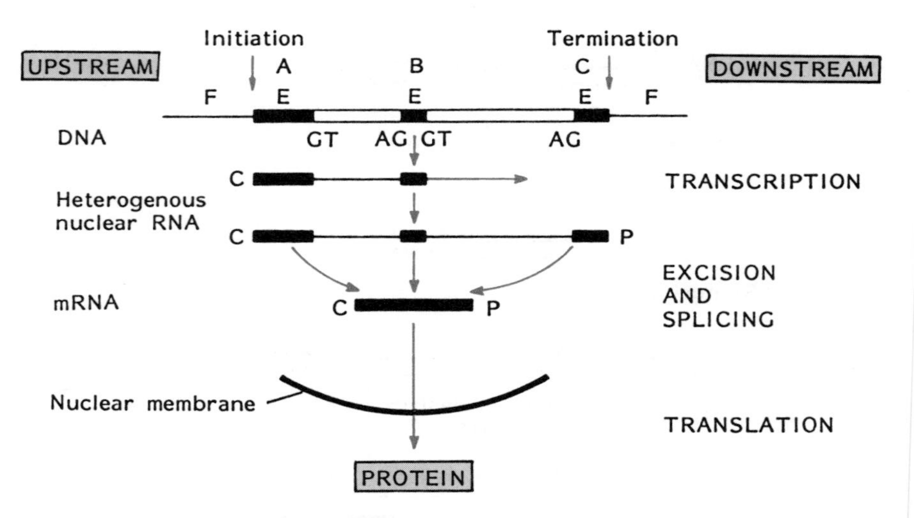

Figure 4.9 Transcription of a gene consisting of introns (I) and exons (E). Both ends are untranslated flanking regions (F). 'C' is the methyl cap and 'P' the poly A tail.

ultimately a so-called **poly A 'tail'** of a hundred or so adenylic acid residues is added to its 3' end. These structures may stabilize the nascent RNA. After splicing the mRNA is transported from the nucleus to the cytoplasm where protein synthesis occurs.

It was shown by Tom Cech that in certain primitive organisms, such as *Tetrahymena*, their RNA contained introns which could be spliced autocatalytically, i.e. with no intermediate other than the RNA. For these discoveries he was awarded the Nobel Prize for Chemistry in 1989. The structure of the tetrahymena RNA intermediates fit remarkably closely to the currently favoured model for eukaryotic gene splicing suggesting that splicing may arise from very early evolutionary origins.

Alternative splicing

In the case of some gene complexes one very large RNA molecule is transcribed. This is then processed in different ways to produce several different RNAs which are then translated into different peptides. This so-called **alternative splicing** may take place in the same tissue but at different stages of development or differentiation, as has now been shown to occur during the development of the thoracic body segments in *Drosophila* and in the synthesis of immunoglobulins in mammals. Alternative splicing may also occur in different tissues. A good example of this is in the processing of a particular RNA (Fig. 4.10) which in thyroid tissue results in the synthesis of calcitonin, involved in controlling serum calcium levels, and in neural tissue in the synthesis of a neuropeptide. Thus the alternative splicing of a precursor RNA to yield different RNAs, and thereby different products, provides a means of increasing a gene's repertoire in both embryonic development and cellular differentiation.

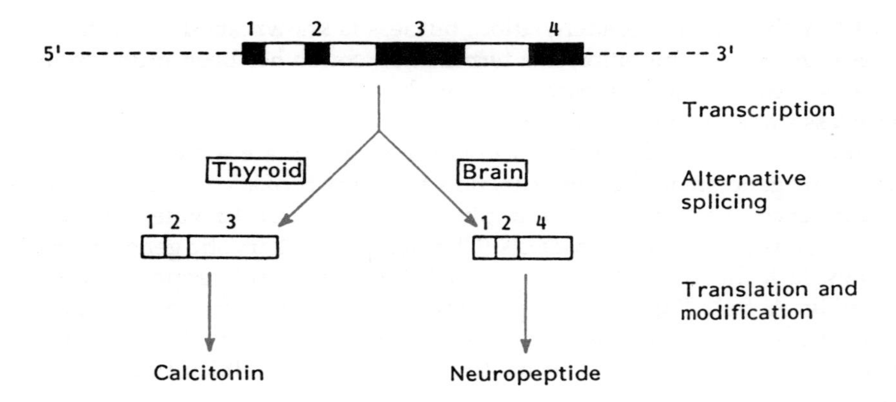

Figure 4.10 Simplified diagram of alternative splicing in the expression of the calcitonin gene. The coding exons (black boxes) are numbered 1 to 4.

Recently Lovell-Badge and colleagues have found that, in adult mouse testis the testis-determining gene, SRY, consists mainly of circular transcripts whereas in the developing genital ridge (in which the gene product presumably mainly acts) a normal product is found. The circular molecules which probably arise through normal splicing mechanisms are not bound to polysomes. If this is designed to stop their translation it remains to be seen how often this mechanism may be used.

Promoter sequences

Upstream from the ATG initiation codon of a gene are a number of non-coding sequences which are referred to as **promoters**. These are believed to be involved in the binding of RNA polymerase and the accurate initiation of transcription, and also possibly in controlling the level of expression of an associated gene(s). About thirty base pairs upstream from the ATG codon is a TATA sequence, sometimes referred to as the Hogness box after its discoverer. Some fifty base pairs further upstream is a CCAAT (or similar) sequence, often referred to as the 'CAT' box, and there may well be other recognition sequences even further upstream which are also necessary for transcription. It would be expected that a mutation in such a promoter sequence could result in reduced transcription of an associated gene and this has in fact been observed in one form of thalassaemia (page 82). It has been known for some time that certain hormones can regulate gene activity and now evidence is accumulating which indicates that these hormones bind to specific DNA sequences. These sequences, which are located to the 5' side of promoters, fall into the general category of so-called **enhancers** which respond to particular proteins in a tissue-specific manner by increasing transcription. Thus the efficient expression of a gene in a particular tissue depends on the right combination and integration of enhancers, promoters and flanking sequences. However, at present there are only vague ideas as to how this integrative activity may occur, though clearly it is the very essence of differential gene activity.

After the termination codon there are also conserved sequences downstream (such as AATAAA) from the gene. They are probably involved in releasing the nascent RNA chain and polyadenylation, but less is known about them than about promoter sequences. A mutation in a termination codon has been found in the case of α-thalassaemia which is associated with an extended transcript which is probably less stable than normal.

From the discussion so far it is now possible to appreciate some of the important differences in gene structure and function between prokaryotes and eukaryotes. These differences have important implications when prokaryotes are used for synthesizing eukaryotic gene products (Chapter 10). Prokaryotic genes do not have introns and lack the enzymes to splice mRNA which is not capped and does not have a poly A tail. Further, the post-translational modifications which many proteins undergo in eukaryotes do not occur in prokaryotes. A number of strategies have been developed to try to overcome these difficulties with varying success.

Identification of coding sequences

As already explained, coding sequences comprise only a small proportion of human DNA and some exons can be less than 100 base pairs long. The methods which have been developed for positional cloning lead to the isolation of cloned stretches of DNA but do little to identify coding sequences. Four main methods are currently in use. The best proven method to date is by the identification of short sequences which have been conserved during evolution. It is likely that exons will have changed their sequence relatively little compared to introns or the DNA between genes. Therefore, a probe made from a cosmid, for example, is hybridized to a Southern blot containing tracks of restriction enzyme digested DNA from many species such as cow, mouse, chick and drosophila (a Zoo blot). Any fragments which cross-hybridize are likely to contain coding sequences and are investigated further. This method, although tedious, has been used to isolate many human genes of different types including dystrophin and the genes causing cystic fibrosis, retinoblastoma and Huntington's chorea.

A second method relies on the fact that the region around the 5' start of a gene and sometimes the first exon contains more sequences of the type 5'CpG3'. These regions can be identified with rare cutter restriction enzymes which have CpG in their restriction site. Identification of a CpG rich island of this sort is regarded as a useful marker for the presence of a gene (Fig. 4.11).

A third method acquiring increasing popularity is exon-trapping or exon-amplification (Church *et al.*, 1994). A vector has been designed so that it contains an intron of the HIV-1 tat gene surrounded by short sections of the surrounding exons. If DNA containing an exon is cloned into a multiple cloning site within the intron all the signals are present to splice that exon into the HIV tat gene product (Fig. 4.12). If the vector is used to transfect cos cells all the material is present for production of mRNA not only of the original size of the HIV tat gene but also longer mRNA containing the spliced in intron. A useful adaptation to the vector is that the multiple cloning site contains a cutting site for the restriction enzyme *Bst*XI. If there is no inserted DNA then the vector can be cut by *Bst*XI before transfection of the cos cells, which will lower the background. Only small amounts of mRNA

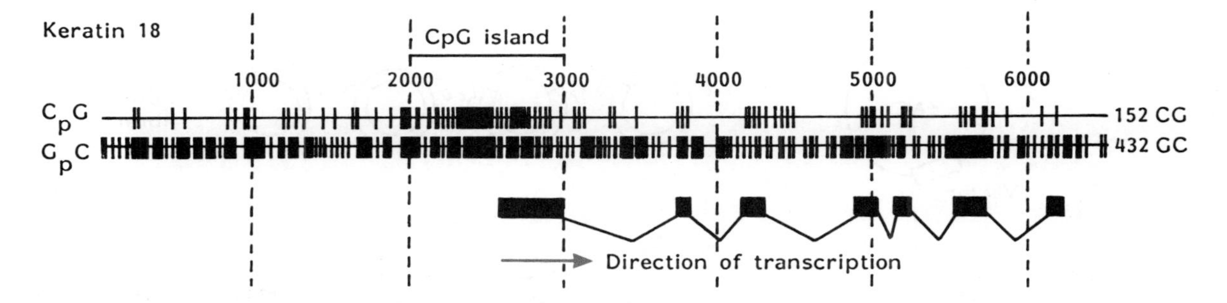

Figure 4.11 A CpG island in the promoter region of a keratin gene. A comparison with GpC shows the comparative scarcity of CpG dinucleotides. (From Wellcome Trust, 1993, with permission.)

are produced but RT–PCR using primers within the flanking exons can be used to amplify the product. The isolated exon can be used directly as a probe.

A fourth method, or set of methods, is based on direct selection of cDNA molecules corresponding to the target sequence. Many variations are possible but in general cosmid or YAC DNA is attached to a solid support such as beads or a filter. Pooled cDNA molecules are incubated with the support, washed and the bound product eluted (Fig. 4.13). Several rounds of enrichment can be carried out. Originally such methods were ruled out because of the small amount of material

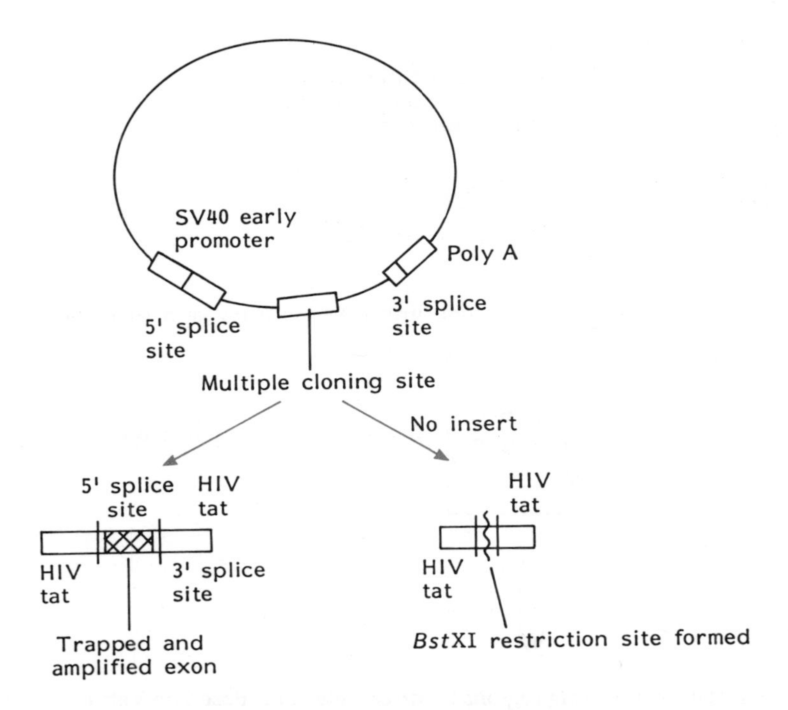

Figure 4.12 pSPL3 vector for exon amplification.

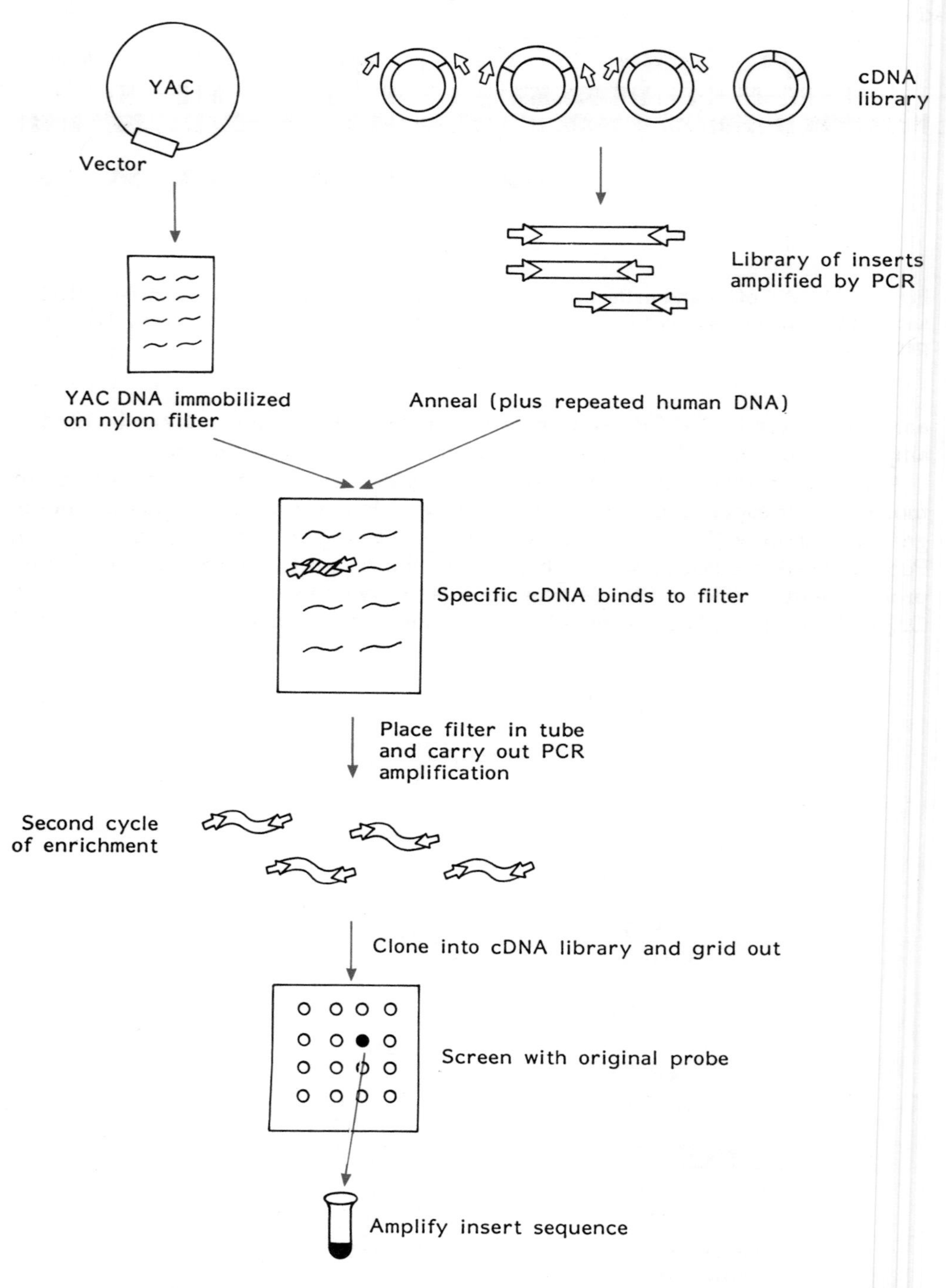

Figure 4.13 Isolation of coding regions by direct selection (Based on Vetrie *et al.*, 1993).

produced. Now, using cDNA libraries and careful choice of primers from the cloning vector, PCR can be used to amplify the small amount of product produced at each stage. This method was used with great effect to isolate the gene for the immunodeficiency, Bruton's agammaglobulinaemia.

Pseudogenes

Pseudogenes are replicas of active genes but do not yield a recognizable gene product. They are one of two types. Some arose by duplication of active genes and then during evolution accumulated various mutations which rendered them inactive. They are a set of genetic quasars and are recognized by DNA homology. The δ-globin gene within the β-globin cluster may be such a pseudogene in the process of creation. It shares strong sequence homology with β-globin but is expressed at much lower levels. There is no obvious physiological necessity for the δ chain and it seems likely that it is accumulating mutations which will eventually render it inactive.

The second class of pseudogenes have apparently arisen by a quite different mechanism. They are often found on different chromosomes from the parent gene and although they share similarities of the coding region they lack introns. They are believed to have arisen through an RNA intermediate, although the details are not known.

The sites of the various genes and pseudogenes in the haemoglobin loci of man and the fine structure of the α- and β-globin genes are illustrated in Fig. 4.14.

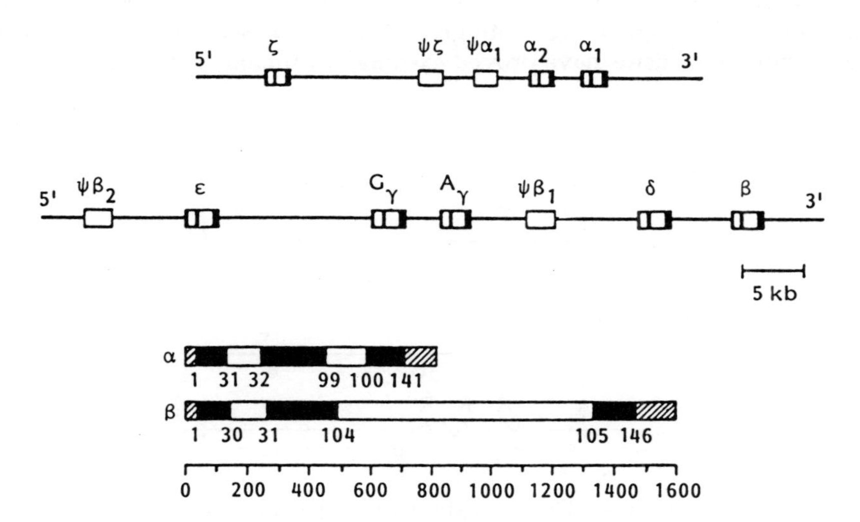

Figure 4.14 Above: the α- and β-globin gene regions on chromosomes 16 and 11 respectively. Pseudogenes are indicated by ψ. Below: Fine structure of the α- and β-globin genes; black boxes represent exons, open boxes represent introns, hatched regions represent untranslated sequences and the numbers represent positions of amino acid codons. The scale is in base pairs.

DNA polymorphisms

Between genes are large stretches of DNA referred to as **intergenic regions** (or 'spacers') with as yet no known function. In fact less than 10% of human DNA is actually involved in protein synthesis. The rest has been referred to as 'junk' DNA to emphasize its apparent redundancy. Within this DNA, as elsewhere in the genome, are variations in nucleotide sequences which have no apparent phenotypic effects on the host organism. These changes in base sequences mean that the restriction site for a particular enzyme may be altered and the fragments so produced by the enzyme will therefore be of different lengths in different individuals. These genotypic changes can be recognized by the different mobilities of the restriction fragments on gel electrophoresis. For example, a restriction site may be lost and thereby a larger DNA fragment will be produced (Fig. 4.15). Such variations occur as frequently as once in every hundred base pairs. Because these variations are relatively frequent in the general population they are referred to as **polymorphisms** and because they are recognized by differences in restriction fragment lengths they are referred to as **restriction fragment length polymorphisms** (RFLP). Each is inherited as a simple Mendelian trait and the pattern of inheritance is so-called co-dominant. Of course, many other base pair variations will occur outside convenient restriction enzyme sites.

Not all variations in restriction fragment lengths reflect variations in nucleotide sequences. Another class of DNA length polymorphisms results from variations in the number of **tandem repeats** between restriction sites. Thus a restriction fragment may be longer in some individuals because the fragment produced by a particular restriction enzyme contains more repeats than in other individuals (Fig. 4.15). The molecular basis of the polymorphism in this case is the variation in the number of tandem repeats. Through linkage with genes for various serious inherited disorders such DNA polymorphisms have proved extremely valuable in carrier detection and

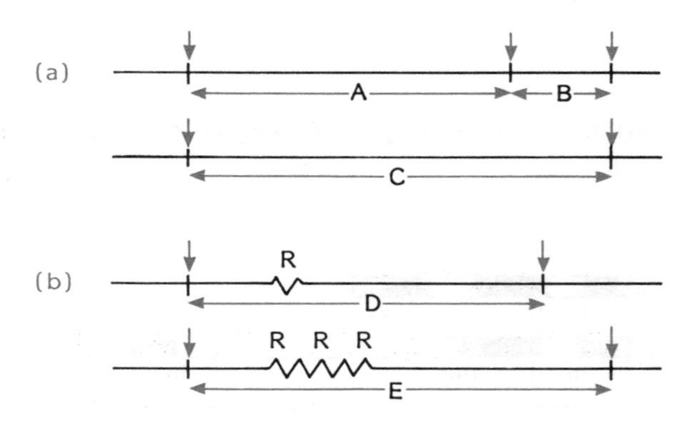

Figure 4.15 DNA length polymorphisms due to: (a) variations in restriction sites, e.g. with loss of a restriction site resulting in a larger DNA fragment (C); (b) variations in the number of repeats (R) between restriction sites. Restriction sites are indicated by vertical arrows and the resultant restriction fragments by letters.

prenatal diagnosis. The most numerous class of tandem repeats, that based on runs of the CA dinucleotide, are distributed virtually at random (Todd, 1993).

Multigene families

An important fact which has emerged has been that certain traits are controlled by several genes with related functions which occur either on the same chromosome (e.g. the α-interferon genes on chromosome 9 and the major histocompatibility gene complex on chromosome 6) or two different chromosomes (e.g. the α-globin and β-globin related gene loci on chromosomes 16 and 11 respectively) or on several different chromosomes (e.g. argininosuccinate synthetase is dispersed over at least eight autosomes and both sex chromosomes). There are several possible ways in which the various members of families of related DNA sequences might become dispersed but much is still conjectural and why such dispersion should have occurred in evolution still remains a mystery. The concerted effort of all members of such gene families is necessary if normal function is to be preserved.

Transposons

In the early 1950s Barbara McClintock first reported on certain genes in maize, responsible for mottling of seed colour, which moved around the genome and were referred to as **controlling elements**. For a long time little attention, or even belief, was afforded to such mobile genetic elements. However, in the last few years it has gradually been accepted that mobile genetic elements, or **transposons**, do occur widely in nature from bacteria and yeast to *Drosophila* (Table 4.1). Some thirty years after her initial discovery of mobile genetic elements, in 1983, Dr McClintock was awarded the Nobel Prize for Medicine for her work.

Transposons possess repeated sequences at either end which are in some way involved in the integration process at a particular point in the DNA molecule. Once inserted they may have several effects. They may produce mutations, a variety of chromosomal rearrangements or, if inserted within a gene, result in complete loss of gene activity. Certainly retroviruses, which cause various neoplasms in animals, become integrated into the host genome (so-called **provirus**) and certain DNA sequences have been found in man which are homologous to retroviral

Table 4.1 Some examples of mobile genetic elements (transposons) in various organisms

Organism	Transposon	Function
Bacteria	Insertion sequences (IS)	–
	Transposons (Tn)	Antibiotic resistance
Yeast	'Cassettes'	Mating types
Maize	Controlling elements	Seed colour
Drosophila	Copia elements, P elements, etc.	Various
Mammals	Provirus (retrovirus sequences)	Cellular oncogenes

oncogenes (page 117). Incidentally, sleeping sickness, which is due to infection by trypanosomes, has proved particularly difficult to control because infection does not confer protection against subsequent reinfection. This is due to the ability of trypanosomes to change their surface antigens and so keep a step ahead of their host's immune system. Evidence now indicates that the ability of trypanosomes to change their surface antigens in this way is due to the action of a transposon.

SUMMARY

- Many hundred gene loci have now been mapped on human chromosomes by various methods, including *in situ* hybridization with gene-specific probes and linkage mapping.
- Almost all genes in eukaryotes are interrupted by intervening sequences or introns.
- During the process of gene transcription a methyl cap and poly A tail are added to the precursor RNA.
- Subsequently, the RNA derived from the introns is excised and the RNA from the exons is spliced together to form the definitive mRNA which is transported to the cytoplasm where protein synthesis occurs.
- Alternative splicing of a precursor RNA to yield different RNAs is an important mechanism in development and cellular differentiation.
- Various DNA sequences, referred to as promoters, which occur upstream from functioning genes (such as the TATA or Hogness box, and the CAT box) are important for transcription. Other upstream sequences (referred to as tissue specific enhancers) determine the correct tissue expression.
- Other genetic structures include pseudogenes, which are inactive replicas of functioning genes, and transposons, which are mobile genetic elements.
- Variations in nucleotide sequences which have no phenotypic effects on the individual are referred to as DNA polymorphisms. Through linkage with genes for serious inherited disorders they are proving useful in carrier detection and prenatal diagnosis.
- With this background to the structure and function of normal genes it is now possible to consider the molecular pathology of some human diseases which have been revealed by recombinant DNA technology.

REFERENCES AND FURTHER READING

Textbooks and review articles

Ballabio A. The rise and fall of positional cloning. *Nature Genetics* 1993; **3**: 277–279

Collins FS. Positional cloning: Let's not call it reverse anymore. *Nature Genetics* 1992; **1**: 3–6

Darnell JE. The processing of RNA. *Sci Amer* 1983; **249**(4): 72–82

McKusick VA, Amberger JS. The Morbid Anatomy of one human genome: chromosomal location of mutations causing disease. *J Med Genet* 1994; **31**: 265–279.

McKusick VA. The defect in Marfan syndrome. *Nature* 1991; **352**: 380
McKusick VA. *Mendelian Inheritance in Man*, 9th edn. London and Baltimore: Johns Hopkins Press, 1992
Todd JA. La carte des microsatellites est arrivee. *Hum Mol Genet* 1992; **1**: 663–666
Tommerup N. Mendelian cytogenetics. Chromosome rearrangements associated with Mendelian disorders. *J Med Genet* 1993; **30**: 713–727
Wise JA. Guides to the heart of the spliceosome. *Science* 1993; **262**: 1978–1979

Research publications

Berk AJ, Sharp PA. Sizing and mapping of early adenovirus mRNAs by gel electrophoresis of S1 endonuclease digested hybrids. *Cell* 1977; **12**: 721–732
Church DM, Stotler CJ, Rutter JL, Murrell JR, Trofatter JA, Buckler AJ. Isolation of genes from complex sources of mammalian genomic DNA using exon amplification. *Nature Genetics* 1994; **6**: 98–105
Davidson N. Proposed banding pattern for human chromosomes. *Cytogenet Cell Genet* 1978; **31**: 313–404
D'Eustachio P, Ruddle FH. Somatic cell genetics and gene families. *Science* 1983; **220**: 919–924
Dietz HC. Marfan syndrome caused by a recurrent de novo missense mutation in the fibrillin gene. *Nature* 1991; **352**: 337–339
Higgs DR, Goodbourne SEY, Lamb J, Clegg JB, Weatherall DJ, Proudfoot NJ. α-Thalassaemia caused by a polyadenylation signal mutation. *Nature* 1983; **306**: 398–400
Jeffreys AJ, Flavell RA. The rabbit β-globin gene contains a large insert in the coding sequence. *Cell* 1977; **12**: 1097–1108
Jeffreys AJ. DNA sequence variants in the Gγ-, Aγ, δ- and β-globin genes of man. *Cell* 1979; **18**: 1–10
Rosenfeld MG, Mermod J-J, Amara SG, Swanson LW, Sawchenko PE, *et al.* Production of a novel neuropeptide encoded by the calcitonin gene via tissue-specific RNA processing. *Nature* 1983; **304**: 129–135
Vetrie D, Vorechovsky I, Sideras P, Holland J, Davies A, Flinter F, Hammarstrom L, Kinnon C, Lerinsky R, Bobrow M, Smith CIE, Bentley DR. The gene involved in X-linked agammaglobulinaemia is a member of the src family of protein-tyrosinekinases. *Nature* 1993; **361**: 226–233
Wyman AR, White R. A highly polymorphic locus in human DNA. *Proc Natl Acad Sci USA* 1980; **77**: 6754–6758

Chapter 5
Molecular pathology of single gene disorders

A **unifactorial** disorder is an inherited condition due to a single gene defect and some 5000 such disorders (Fig. 5.1) have now been recognized (McKusick, 1992). Most are very rare although there are some notable exceptions, such as cystic fibrosis, with an incidence in Caucasians of about one in 2000 total live births, and Duchenne muscular dystrophy, with an incidence of about one in 3000 male live births. Almost all unifactorial disorders are serious and for very few is there any really effective treatment.

A unifactorial trait which is determined by a gene on an autosome is said to be inherited as an **autosomal trait** and may be dominant or recessive. A unifactorial trait which is determined by a gene on the X chromosome (there are no known genes for any serious diseases on the Y chromosome) is said to be **X-linked** and, with very few exceptions, these are inherited as recessive traits. To put such disorders into perspective the total incidence per 1000 births is about 7 for autosomal dominant disorders, 2.5 for autosomal recessive disorders, and 0.5 for X-linked recessive disorders. Some of these disorders (usually autosomal recessives) will present at birth but many will not manifest themselves until later in life. Congenital abnormalities occur in about one in every fifty newborns and about a fifth of these abnormalities are due to unifactorial disorders.

Since unifactorial disorders obey Mendelian laws of inheritance they are sometimes also referred to as **Mendelian disorders**. Details of the modes of inheritance can be found in several introductory textbooks of genetics, but the essentials may be summarized as follows. A person with a rare **autosomal dominant** disorder possesses both the abnormal (mutant) dominant gene which causes the disorder as well as its normal homologue or allele. Affected individuals are thus said to be heterozygous and half their gametes will carry the normal gene and half the mutant gene. Therefore on average half the offspring of any affected individual will also be affected. The less serious autosomal dominant disorders can often be traced through several generations of a family (Fig. 5.2), as can those like Huntington's chorea where the onset may be after the affected individual has had children. However, in the case of other serious diseases affected individuals may not live long enough to have a family, or may survive but have reduced fertility, and here the disease ultimately becomes extinct in a family and is maintained in the population by new mutations occurring in other families. Examples of autosomal dominant disorders include milder forms of brittle bone disease (osteogenesis imperfecta), myotonic dystrophy and Huntington's chorea.

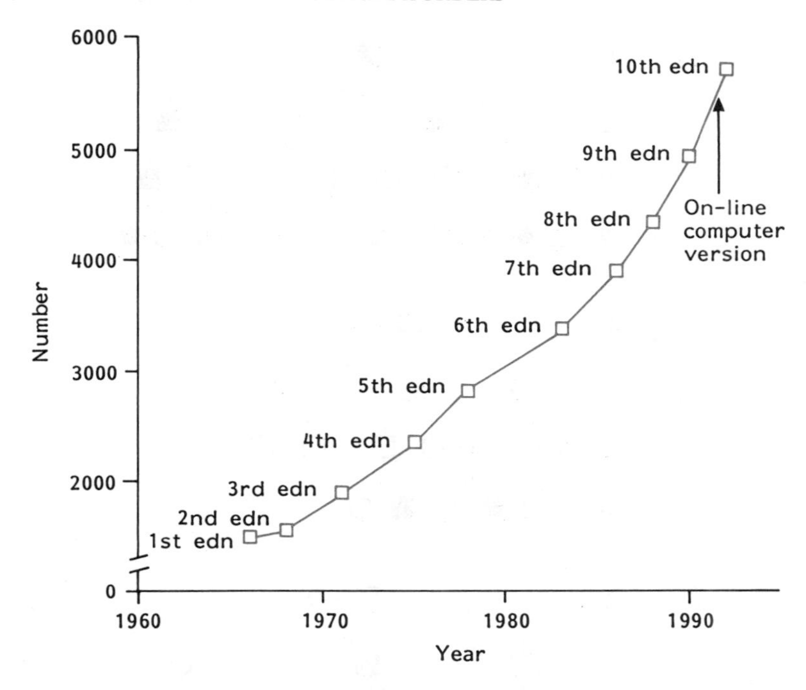

Figure 5.1 Numbers of confirmed gene loci given in McKusick's catalogues from 1966 to 1992. Including unconfirmed loci, the total number of disorders listed by 1992 was 5710.

Autosomal recessive disorders, on the other hand, are only manifest when the mutant gene is present in double dose and such affected individuals are therefore said to be homozygous for the gene in question. Heterozygotes are unaffected and all their offspring will be unaffected unless by chance two heterozygotes happen to marry. In the latter case each parental gamete will carry either the mutant gene or the normal gene. The chance of a mutant gene from each parent combining is one in four, which is therefore the chance of two heterozygous parents having an affected child. The rarer the recessive gene the more likely the parents are to be related (consanguineous) because they are more likely to have both inherited the same rare recessive gene from a common ancestor. Should an affected homozygous individual survive and marry, unless the spouse is a heterozygote which is unlikely because most of these diseases are rare, then all the offspring will be heterozygotes and therefore unaffected. Thus in autosomal recessive disorders it is not possible to trace the disease through several generations and all affected individuals in a family are in one sibship, i.e. they are brothers and sisters (Fig. 5.2). Most autosomal recessive disorders present at birth and include many metabolic diseases, cystic fibrosis, certain types of deaf-mutism and congenital blindness, sickle cell anaemia and thalassaemia major.

X-linked recessive disorders are not usually expressed in the heterozygous female because she carries the normal (dominant) allele on her other X chromosome. However, in the male a mutant gene carried on his single X chromosome is always manifest because there is no normal allele to counteract the

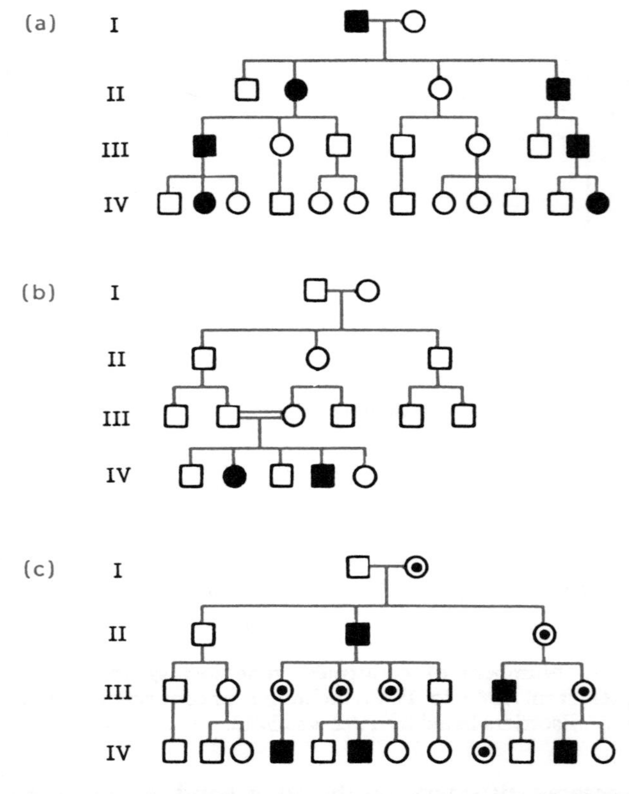

Figure 5.2 Pedigree pattern of an: (a) autosomal dominant trait, (b) autosomal recessive trait, (c) X-linked recessive trait.
Pedigree symbols: □○ Normal male and female
■● Affected male and female
□═○ Consanguineous mating
⊙ Female heterozygous carrier of an X-linked disorder.

effects of the mutant gene as there is in the heterozygous female. Since the carrier female may transmit either her X chromosome bearing the mutant gene or her X chromosome bearing the normal allele to any of her offspring, on average half her sons will be affected and half her daughters will be carriers (Fig. 5.2). In serious X-linked disorders, where affected males do not have children, these are transmitted entirely by healthy female carriers. However, in the case of less serious X-linked disorders, where affected males can survive to have children, then since a father transmits his X chromosome to all his daughters but to none of his sons, all the daughters of an affected male will be carriers, but all his sons will be unaffected. Examples of X-linked recessive disorders include all the various diseases given in Fig. 4.2 (page 58), including Duchenne muscular dystrophy and haemophilia.

Over the last decade there has been an explosion in the number of disease-causing genes which have been isolated. The almost daily new discoveries make it impossible to provide a comprehensive list. Roughly speaking the discoveries divide into two types. The first has been the isolation of genes, previously unknown, by the positional cloning approach already described. The second has been the isolation of genes which were already known because of their biochemical properties (Ballabio, 1993). Prime examples have been the factor VIII and IX genes which cause haemophilia A and B, phenylalanine hydroxylase which causes phenylketonuria (PKU) and genes for many metabolic and lysosomal enzymes. The most useful technology in this second case has been to work back from the sequence of amino acids in the protein to the gene. If even a small amount of purified protein can be isolated modern protein sequencers can determine some amino acid sequence, particularly at the amino-terminal. Because the genetic code is known for each amino acid it is possible to predict the underlying gene sequence. Unfortunately, because there is more than one codon for most of the amino acids it is not possible to predict each base accurately. Oligonucleotides are then synthesized for all possible combinations and the oligonucleotide mixture used to probe a library of clones made from RNA from a suitable tissue.

HAEMOGLOBINOPATHIES

The globin genes were among the first to be isolated and because of that and also because of the large number of genetic diseases caused by alterations in haemoglobin the haemogolobinopathies have been extensively studied. The observations in them have formed the foundation for all other single gene disorders studied subsequently and it is a rare lesson that we cannot learn first from the study of haemoglobinopathies. We will, therefore, review them in detail before considering some other important disorders.

More has been learned by recombinant DNA technology of the molecular pathology of the haemoglobinopathies than any other group of disorders. There are perhaps two reasons for this. Firstly, these disorders have always attracted considerable interest from medical scientists because of their clinical importance. On a global scale they cause far more ill health than any other group of inherited conditions. About a quarter of a million severely affected individuals are born each year with one or other of these disorders (Weatherall, 1993). Secondly, their study has been made easier by the fact that tissues in which gene expression is limited to the synthesis of globin chains of haemoglobin can be obtained relatively easily (peripheral blood reticulocytes).

The haemoglobin molecule is composed of four subunits, each of which consists of an iron-containing haem portion and a polypeptide chain (globin). Normally each molecule has two polypeptide chains of one sort and two of another. The different sorts of polypeptide chains are referred to by Greek letters: alpha α, beta β, gamma γ, delta δ, epsilon ε and zeta ζ. On chromosome 16 is situated the alpha-like globin gene complex which encodes the various alpha-like globin chains and consists of two active α genes and a ζ gene. On chromosome 11 is situated the beta-like globin gene complex which codes the various beta-like globin chains and consists of an ε gene, two different γ genes (Gγ, Aγ), a δ gene and a β gene (Fig. 4.14). The ζ and the ε genes are only active in embryo, the γ genes normally only in the fetus and the β and the δ genes only in the adult. The various globin chains and their respective haemoglobins are summarized in Table 5.1.

Table 5.1 Human haemoglobins synthesized at different stages of development and in the normal healthy adult

Stage in development	Haemoglobin	Structure	Proportion (%) in normal adult
Embryonic	Gower I	$\zeta_2\varepsilon_2$	–
	Gower II	$\alpha_2\varepsilon_2$	–
	Portland I	$\zeta_2\gamma_2$	–
Fetal	F	$\alpha_2\gamma_2$	<1
Adult	A	$\alpha_2\beta_2$	97–98
	A_2	$\alpha_2\delta_2$	2–3

Structural haemoglobin variants

The haemoglobinopathies can be divided into two major groups. The first group are those in which there is a **structural** alteration in one of the globin peptide chains. Such alterations result from point mutations which may occur in any of the 141 amino acids in the α chain, the 146 amino acids of the β chain or less rarely in the γ or δ chains. The classic example in this group is sickle cell anaemia, a disease which occurs mainly in Africa, parts of the Mediterranean area and India. Affected individuals suffer from a very severe haemolytic anaemia and death often occurs in childhood, though occasionally individuals may survive into adulthood. It is due to a point mutation (transversion) at the sixth codon from the 5' end of the β-globin gene which results in the replacement of glutamic acid by valine:

```
                     glutamic acid
β^A  ..........  CCT      GAG       GAG  ..........
                           ↓
β^S  ..........  CCT      GTG       GAG  ..........
                         valine
```

Sickle cell anaemia is a recessive disorder and affected individuals are homozygous for the mutant gene.

Some other examples of structural alterations in globin chains are given in Table 5.2. In fact over 300 such disorders have now been described in each of which a specific point mutation has been identified and which results in the substitution of a different amino acid at a particular point in the globin chain.

The molecular basis of these disorders has gradually been worked out over the last 25 years or so, largely as a result of applying standard biochemical techniques such as electrophoresis, chromatography and amino acid sequencing. Though recombinant DNA technology has played relatively little part in the elucidation of these disorders, it has certainly opened up new avenues for diagnosing them prenatally (page 146). However, in the second group of haemoglobinopathies the picture has been quite different. In this second group the basic abnormality is a defect in globin chain **synthesis** rather than a structural abnormality in the globin chain itself. This second group constitute the **thalassaemias** and here recombinant

Table 5.2 Structural variants of haemoglobin

Type of mutation	Examples	Chain/residue(s)/alteration
Point (over 200 variants)	Hb S Hb C Hb E	beta, 6 glu to val beta, 6 glu to lys beta, 26 glu to lys
Deletion (shortened chain)	Hb Freiburg Hb Lyon Hb Leiden Hb Gun Hill	beta, 23 to 0 beta, 17–18 to 0 beta, 6 or 7 to 0 beta, 92–96 or 93–97 to 0
Insertion (elongated chain)	Hb Grady	alpha, 116–118 (glu, phe, thr) duplicated
Frameshift (insertion/deletion) of other than multiples of 3 base pairs	Hb Tak Hb Cranston Hb Wayne	*beta, +11 residues, loss of termination codon, insertion of 2 base pairs in codon 146/147 *alpha, +5 residues, due to loss of termination codon by single base pair deletion in codon 138/139
Chain termination	Hb Constant Spring Hb McKees Rock	*alpha, +31 residues, point mutation in termination codon *beta, −2 residues, point mutation in 145, generating premature termination codon
Fusion chain (unequal crossing-over)	Hb Lepore/anti-Lepore Hb Kenya/anti-Kenya	non-alpha, delta-like residues at N-terminal end and beta-like residues at C-terminal end, vice versa, respectively non-alpha, gamma-like residues at N-terminal end and beta-like residues at C-terminal end, vice versa, respectively

* Residues are either added (+) or lost (−).

DNA technology has been extremely valuable in unravelling the complex molecular pathology of this diverse group of diseases. The whole gamut of the genetic engineer's technology has been used, including the screening of genomic libraries with cDNA probes, the isolation, characterization and sequencing of globin genes, and *in vitro* translation and protein synthesis. A recent volume contains excellent up-to-date information on the full range of mutations (see Baillière's *Clinical Haematology. The Haemoglobinopathies*).

The thalassaemias are so named after the Greek word 'thalassa' (θαλασσα) meaning 'sea', since many forms of this disease occur around the Mediterranean, though in fact they also occur very frequently in the Middle East, India, Burma and South-East Asia. They are the commonest single gene disorders in the world population and produce massive public health problems in many countries.

They are conveniently classified according to which particular globin chain is affected, the most clearly defined being the α-, β- and $\delta\beta$-thalassaemias. The molecular defects and clinical features of the more important forms of thalassaemia are summarized in Table 5.3. The picture, however, may be further complicated because these disorders frequently occur in populations in which there is also a high prevalence of inherited structural haemoglobin variants. It is therefore not uncommon for a patient to have thalassaemia as well as an abnormal haemoglobin, or two different haemoglobin variants, or even two different types of thalassaemia! The clinical picture can therefore be confusing but makes better sense when the molecular basis of these diseases is taken into account.

Table 5.3 Molecular defects and clinical features of various forms of thalassaemia. Deletion of three α genes (referred to as α^0/α^+) produces haemoglobin H disease

Disease	Molecular defect	Clinical	
		Homozygote	Heterozygote
α-Thalassaemias			
α^0	• Deletion of two linked α genes	Hydrops fetalis with Hb Bart's	Mild anaemia
α^+	• Deletion of one of the pair of linked α genes	Mild anaemia	Normal
	• Mutation of stop codon (e.g. Hb Constant Spring)	Moderate anaemia	Normal
	• Various other mutations	Variable, mild anaemia or Hb H disease in some cases	Normal or mild anaemia
β-Thalassaemias			
β^0	• Partial deletion of β gene	Anaemia	Mild anaemia
	• Mutations resulting in premature stop codon and frameshifts	Anaemia	Mild anaemia
	• Splice junction mutations		
β^+	• Mutations affecting splicing	Anaemia	Mild anaemia
	• Promoter sequence mutations		
$\delta\beta$-Thalassaemias			
	• Various deletions and fusions	Severe anaemia (Hb Lepore)	Mild anaemia
		Mild anaemia ($\delta\beta$-thal.)	Very mild anaemia
		Very mild anaemia (HPFH)	Normal

α-Thalassaemias

The α-thalassaemias occur most commonly in the East. They can be divided into two forms: a severe form referred to as α^0-thalassaemia and a milder form referred to as α^+-thalassaemia. Studies of DNA from patients have revealed the underlying defects in these disorders.

The α^0-thalassaemias have been shown to result from deletions of **both** α-globin genes on a chromosome, whereas the α^+-thalassaemias result from either deletions of **one** of the two linked α-globin genes on a chromosome or less frequently from various non-deletion defects of the α gene which result in reduced synthesis of α chains (Fig. 5.3).

If all four α genes are missing then an excess of γ chains is produced by the fetus which form γ_4 tetramers or haemoglobin Bart's. This is incompatible with survival, the affected fetus being stillborn and grossly oedematous (hydrops fetalis). A deletion of three α genes is not so serious and affected individuals can survive into adulthood with a variable degree of anaemia resulting from an excess of β chains which form β_4 tetramers or haemoglobin H. Individuals with deletions of two α genes (as occurs in α^0 heterozygotes and α^+ homozygotes) or one α gene (as occurs in α^+ heterozygotes) are usually healthy or only have a very mild anaemia. Possible mating types between α^0 heterozygotes and α^+ heterozygotes and the resultant offspring are summarized in Fig. 5.4.

Thus if two α^0 heterozygotes marry, on average one in four of their children will be α^0 homozygotes and die *in utero* from hydrops fetalis with haemoglobin Bart's (Fig. 5.4a). If an α^0 heterozygote marries an α^+ heterozygote then on average one in four of their children will have haemoglobin H disease (Fig. 5.4b). Finally, if two α^+ heterozygotes marry then none of their children will have a severe anaemia (Fig. 5.4c).

Occasionally α^+ thalassaemia may result from a variety of other molecular defects apart from deletions, but the details cannot be considered here. There is, however, one particular defect of interest because of the general principle which it illustrates. This is the so-called haemoglobin Constant Spring, after the town in the United States, the home of the original patient with this abnormality. This results from a mutation which eliminates the normal stop codon in the α gene with the result that the α chain is abnormally long and unstable and results in the clinical picture of α^+-thalassaemia.

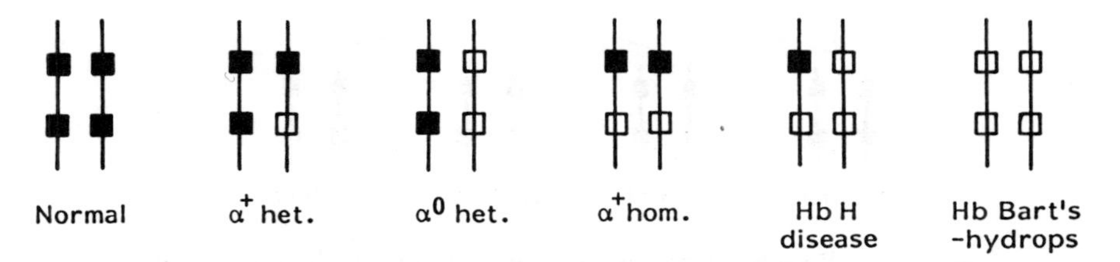

Figure 5.3 Deletions (□) of α genes (■) which occur in the α-thalassaemias.

HbH disease and mental retardation

In a very few cases mild α-thalassaemia has been found associated with a particularly severe form of mental retardation. Molecular genetic studies have shown that these cases divide into two, although they are clinically indistinguishable. The α-globin gene cluster lies very close to the end of the short arm of chromosome 16. In one type the α-thalassaemia is caused by a deletion which extends from the α-globin genes right through to the sequences defining the

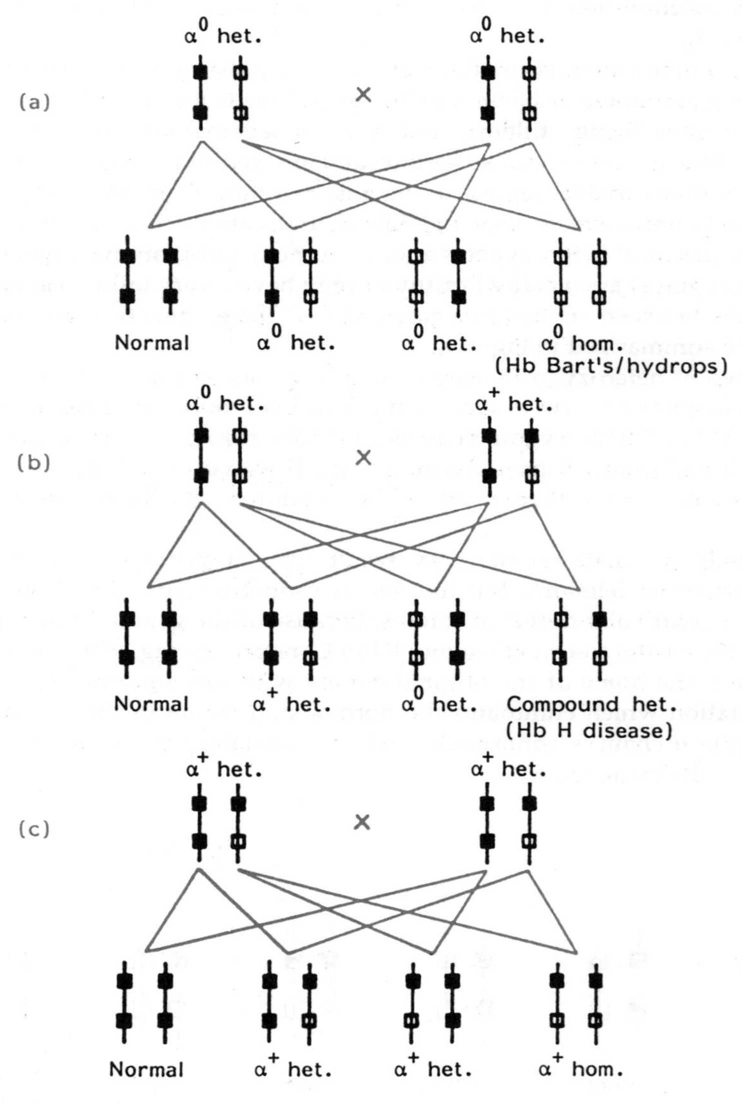

Figure 5.4 Possible matings between α^0 and α^+ heterozygotes and the resultant offspring. □ indicates a deleted α-globin gene.

telomere of the short arm of chromosome 16. In the second type no alterations of the α-globin gene cluster could be found and gradually examples accumulated in which brothers or maternal uncles were also affected. It is now believed that the disorder is caused by a trans acting factor, affecting globin expression coded on the X chromosome.

β-Thalassaemias

The β-thalassaemias can also be divided into two groups: β^+ and β^0, depending respectively on whether there is a reduction or absence of β-chain synthesis. The clinical features vary greatly. In some cases, referred to as **thalassaemia major** or Cooley's anaemia, they present in the first year of life with failure to thrive and a very severe anaemia. In an effort to compensate for the anaemia the bone marrow expands producing deformities of the face and skull. Affected children have recurrent infections and without adequate blood transfusions will eventually die. However, heterozygous carriers of the β-thalassaemia trait are symptomless, and this is sometimes referred to as **thalassaemia minor** or the thalassaemia trait.

The effects of various mutations on β-globin chain synthesis have been investigated by a number of techniques, including analysing RNA synthesis in erythroid cells from patients and studying the expression of the cloned mutant gene after its introduction into cultured cells using a suitable vector such as SV40. Recombinant DNA studies have now revealed many different mutations within and around the β-globin gene which may produce the various forms of β-thalassaemia (Table 5.3). Firstly, there may be defective synthesis of mRNA as a result of nonsense (premature termination) or frameshift mutations within an exon, or mutations in a promoter sequence, or very rarely a deletion of the β-globin gene. Secondly, there may be defective splicing which then results in an altered mRNA which is transcribed less efficiently. Such defective splicing can occur in a number of different ways with different mutations: the normal splice site may be eradicated, or a new alternative splice site may be created, or a pre-existing cryptic splice site in an intron or exon which is not normally used may be activated. The activation of cryptic splice sites may result in the formation of a larger exon or even an extra exon within an existing intron. In all these cases defective splicing results in the formation of an aberrant mRNA.

$\delta\beta$-Thalassaemias

These thalassaemias are due to reduced synthesis of both β and δ chains and are much less common than the α- or β-thalassaemias. They are worthy of special mention because their study has revealed yet another type of defect at the molecular level. The clinical features in affected homozygotes vary considerably from severe anaemia in some cases to a very mild anaemia in what is referred to as **hereditary persistence of fetal haemoglobin** (HPFH). Individuals with HPFH have 100% fetal haemoglobin due to various deletions involving the δ and β-globin genes which is compensated for by the synthesis of γ chains of haemoglobin F (Fig. 5.5).

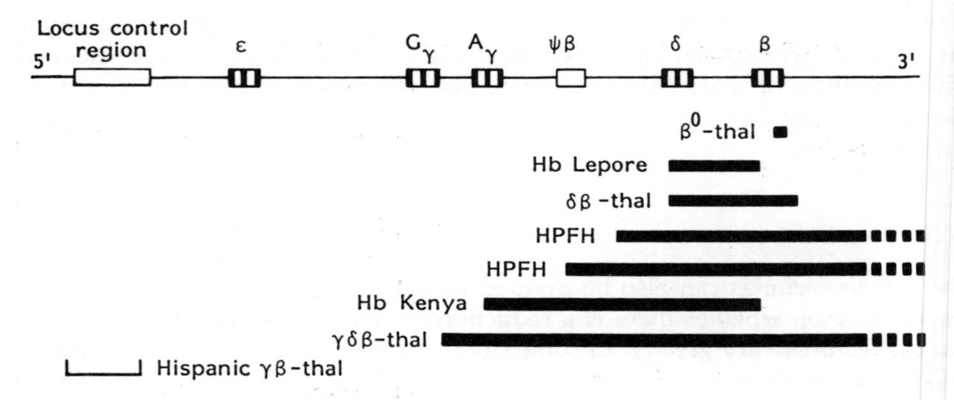

Figure 5.5 Some of the deletions in the β-globin gene region which cause various β-thalassaemia syndromes.

Experiments involving transfer of increasingly long segments of DNA into mouse embryos (transgenic mice) have revealed the need for a stretch of DNA, called a locus control region or LCR, upstream from the structural genes, in order to regulate the expression of the gene cluster correctly. A Hispanic form of $\gamma\beta$-thalassaemia has been found resulting from just a deletion within this region. It is significant that the order of the genes along the chromosome is the same as the order of expression during development (this has been observed for other closely regulated gene clusters such as homeobox genes) and various models for their regulation involving the LCR have been proposed (Fig. 5.6).

However, while a dozen or so different deletions of the β-globin gene region have now been described in patients with this disease, other cases are believed to result from mispairing during meiosis with resultant unequal crossing-over between the β and δ-globin gene loci. This results in the production of $\delta\beta$ fusion gene (Lepore gene) on one chromosome and a $\beta\delta$ fusion gene (anti-Lepore gene) on the other chromosome (Fig. 5.7). Haemoglobin Lepore is named after the family in which it was first described. The Lepore globin chains are synthesized inefficiently and result in thalassaemia. However, unequal crossing-over with the formation of fusion genes is not unique to thalassaemia. The most common form of

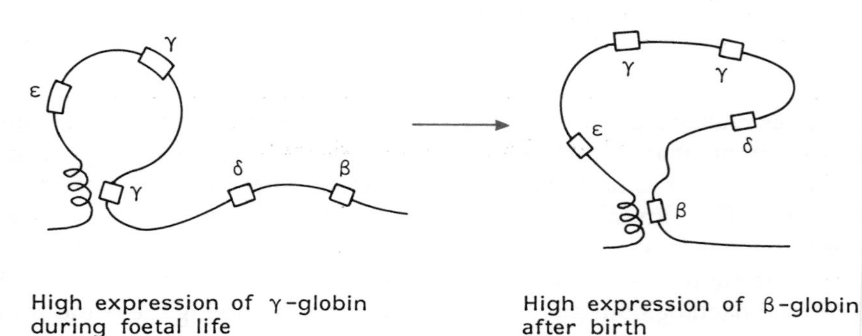

Figure 5.6 Proposed model for LCR control of globin gene regulation.

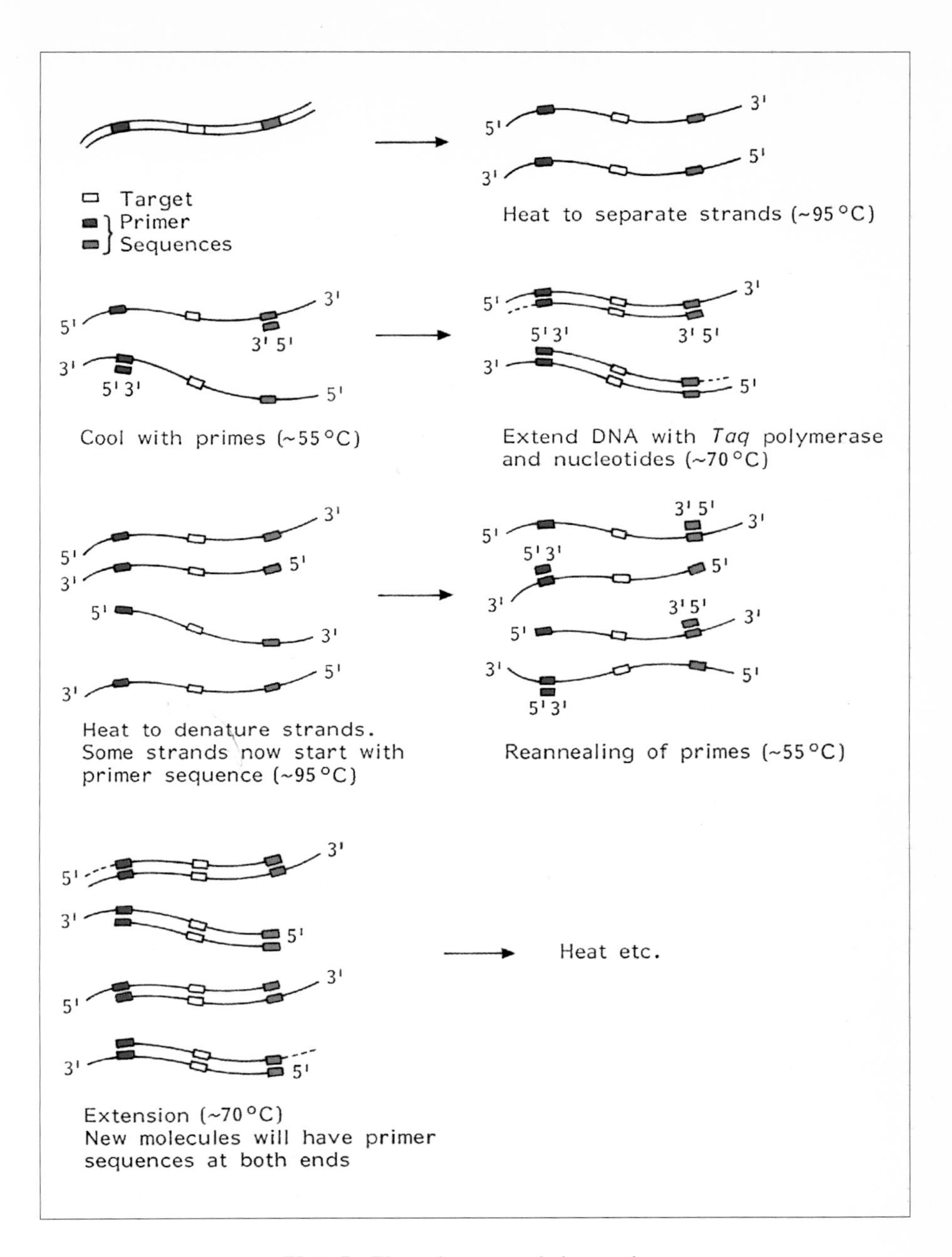

Plate I The polymerase chain reaction.

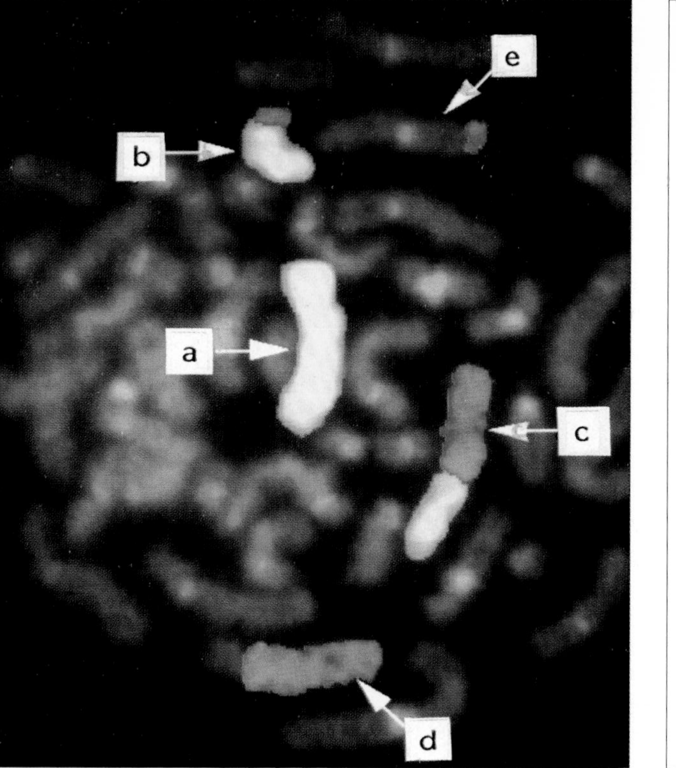

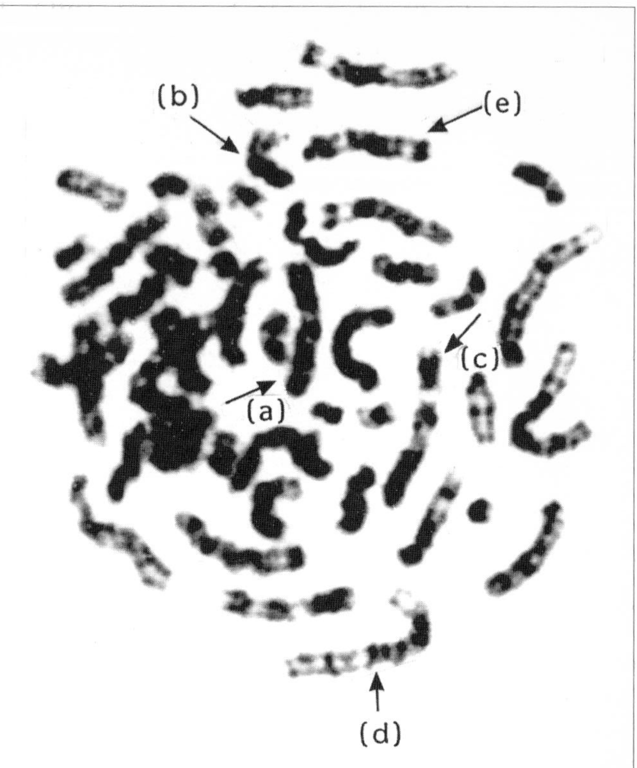

Plate II Giemsa banded metaphase from the bone marrow of a child with acute lymphoblastic leukaemia containing the recurrent translocation t(4;11)(q21;q23) and a translocation involving the second chromosome 11, t(3;11)(q25;q23) (right). The cell has been destained and hybridized with chromosome 4 and chromosome 11 specific whole chromosome paints. Chromosome 4 = yellow (FITC labelled). Chromosome 11 = red (biotin labelled and detected with avidin Texas Red). a = normal chromosome 4 (yellow). b = derivative chromosome 4, t(4;11). c = derivative chromosome 11, t(4;11). d = derivative chromosome 11, t(3;11). e = derivative chromosome 3, t(3;11).

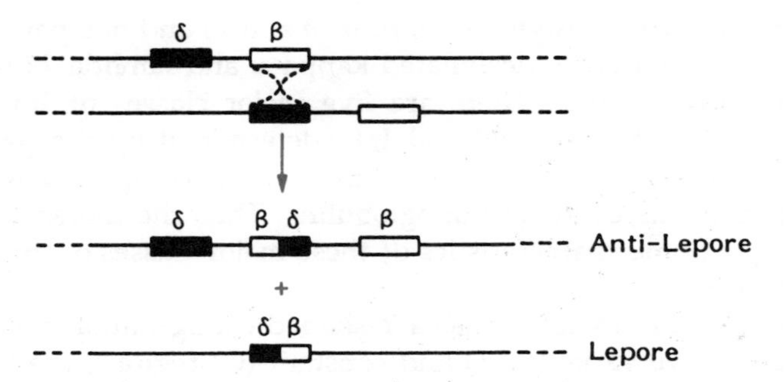

Figure 5.7 Unequal crossing-over between the β- and δ-globin genes.

hereditary motor and sensory neuropathy type I is caused by a duplication of 1.5 Mb (one and a half million base pairs) of DNA arising from just this mechanism. Within the duplication is the gene for a peripheral myelin protein (PMP-22) and the demyelination found in this disorder is believed to arise from over-expression of this gene.

THE IMMUNE SYSTEM

Another field in which recombinant DNA technology has provided detailed information of molecular pathology is in the immune system. The immune system consists of a number of entities: immunoglobulin (antibody) synthesis, cellular immunity, the complement system and the HLA complex. Many of the genes in these different systems have now been cloned and characterized and a great deal of detailed information has been gained. However, it would not be appropriate to discuss all these findings here, but there is one matter which does merit special consideration; that is the way in which antibody diversity is generated, for until the advent of recombinant DNA technology very little was known about this.

The problem posed is how to account for the vast array of antibodies which can be produced in response to the almost limitless number of different foreign antigens to which an individual might be exposed during his or her lifetime. The most facile explanation suggested that each different type of antibody was coded by a different gene, but this would have meant that a significant proportion of the entire genome would be committed to this one function alone, and there was no evidence for this. The answer in fact lies in **somatic recombination**, whereby different parts of the immunoglobulin molecule are put together by various combinations of genes which code for its constituent parts. This idea was first proposed 30 years ago by Dreyer and Bennett, and therefore is sometimes referred to as the Dreyer–Bennett hypothesis, but there was little evidence for it until DNA studies showed that the idea was essentially correct.

The immunoglobulins (Ig) are a class of serum proteins synthesized and secreted by plasma cells derived from B lymphocytes present in lymphoid tissues. Each immunoglobulin molecule is made up of four chains: one pair of heavy chains (of which there are five types designated γ, μ, α, δ and ε) and one pair of light chains (of which there are two types designated kappa κ and lambda λ) which are held together by hydrogen bonds. There are five major classes of immunoglobulins referred to as IgG, IgM, IgA, IgD and IgE, determined by the particular heavy chain present (γ, μ, α, δ and ε respectively). The two types of light chains are common to all five classes of immunoglobulins. Thus the molecular formula for IgG is $\gamma_2\kappa_2$ or $\gamma_2\lambda_2$. Some characteristics of these major classes of immunoglobulins are given in Table 5.4.

Each immunoglobulin molecule has a Y-shaped configuration and both the light and heavy chains have variable (V) and constant (C) regions, the latter being the same in every antibody of a given type, and in the case of heavy chains determines its effector function, such as cell binding, etc. The variable regions are responsible for antigen recognition and binding and can be further subdivided into seven regions: four (framework regions numbered 1 to 4) which vary only slightly from one antibody molecule to another, and three (hypervariable regions numbered I to III) which are very much more variable (Fig. 5.8).

Most of the DNA studies which have revealed how antibody diversity is generated have been carried out in the mouse, but the general picture is similar in the human and only differs in some of the detail. Each light chain is coded for by three separate genes: a C gene for the constant region, a J (for joining) gene mainly for framework region 4 and a V gene for the rest of the variable region. Each heavy chain, on the other hand, is coded for by four separate genes: a C gene for the constant region, a J gene for framework region 4, an additional D gene for hypervariable region III and a V gene for the rest of the variable region (Fig. 5.9).

Cloning and sequencing of these various genes has shown that in the germ line there are a hundred or so alternative V genes, each separated by a short intron from a leader (L) segment. The latter is probably involved in the transport of the antibody molecule through the cell membrane, after which it is cleaved away. Further, there are five or so J genes, and in the case of heavy chains ten or more D genes. The important point is that members of each family of V, D and J genes combine with each other at random by somatic recombination during lymphocyte differentiation. Thus we can begin to see how antibody diversity is generated. If it is **assumed** that in a light chain there are 150 V genes and 5 J genes, the potential

Table 5.4 Some characteristics of the major classes of immunoglobulins

Class	Molecular weight	Serum concentration mg/ml	Antibody activity	Placental transfer
IgG	160 000	12.0	To bacteria and viruses (most antibodies)	+
IgM	900 000	1.0	To protein antigens (some blood groups and bacteria)	–
IgA	170 000+	2.5	In external secretions (local immunity)	?+
IgD	180 000	0.03	?	–
IgE	200 000	Trace	In allergic reactions	–

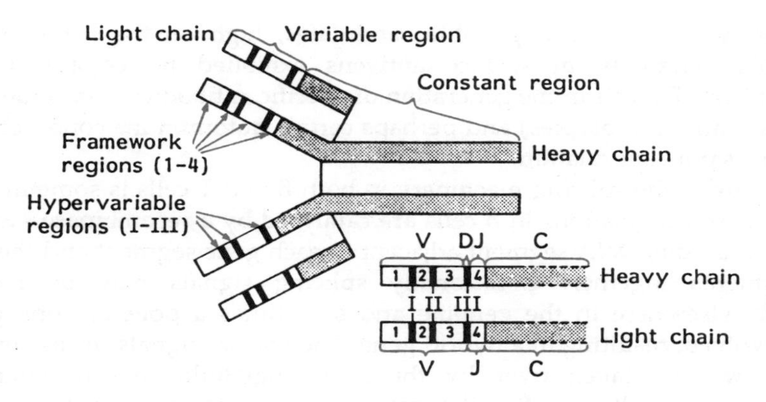

Figure 5.8 The structure of the immunoglobulin molecule.

diversity would be 150 × 5, or 750. To this can also be added variability due to alternative joining sites ('recombinational flexibility') between the different genetic components (Leder, 1982). If this increases the variability, say, tenfold, the potential diversity then becomes 7500. Variation in the heavy chain component must also be included. If it is **assumed** in the heavy chain that there are 80 V genes, 6 J genes and, say, 50 D genes with the same amount of recombinational flexibility at each of the **two** joining sites, then the total variation in possible antibodies becomes 7500 × 80 × 6 × 50 × 100, or 18 billion! However, even this does not exhaust the repertoire because there is also the possibility of somatic mutations occurring in the variable region DNA of developing lymphocytes. The diversity of the T cell receptors is created in a similar way.

Some time has been spent on the generation of antibody diversity because it illustrates the general principle of the enormous potential for variation which can occur through gene shuffling. As Leder (1982), a pioneer in this field, has pointed out, this is likely to be the way in which diversity is produced in some other gene

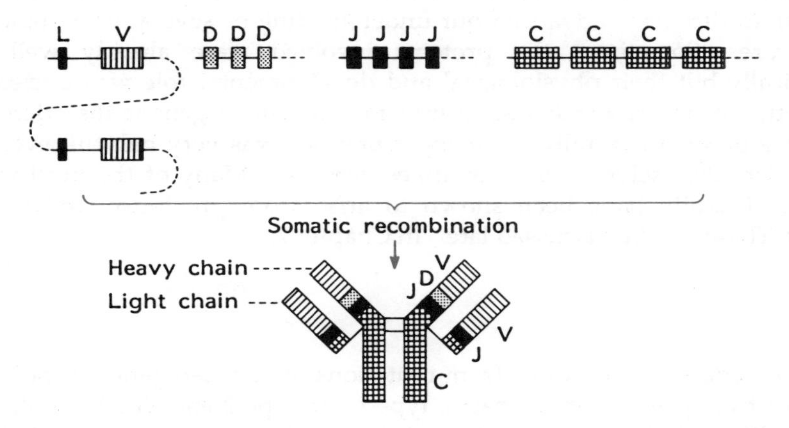

Figure 5.9 Generation of antibody diversity from somatic recombination between members of gene families coding for different parts of the molecule.

systems as well. Of several possible candidates, high on the list must be the considerable variability in surface antigens exhibited by certain pathogenic microorganisms. Defects in the generation of specific antibodies may underlie some inherited sensitivities (atopies) and perhaps certain autoimmune conditions, but at present this is pure speculation.

Unfortunately, the splicing machinery in both B and T cells is sometimes prone to error. The recombinations in B cells are catalysed by a recombinase system that recognizes a unique DNA segment adjacent to each gene segment and then splices those segments together. Occasionally splicing signals may be recognized erroneously elsewhere in the genome and sometimes a potential oncogene can become involved, resulting in a hybrid gene. The normal signals for its control and regulation will be taken over by the immunoglobulin portion and altered expression may result in a B cell tumour. An important example is found in Burkitt's lymphoma in which the c-*myc* oncogene is translocated from chromosome 8 to any one of the immunoglobulin chains, i.e. the heavy chain on chromosome 14 or the light chains on chromosome 2 or 22. In follicular lymphomas there is a frequent translocation resulting in the fusion of the immunoglobulin gene to a region of chromosome 18. By walking from the previously well defined immunoglobulin gene it was possible to clone the gene involved on chromosome 18 which was subsequently named *bcl*-2 and was the first member of an important group of genes regulating programmed cell death. It is hardly surprising that the incorrect regulation of such a gene should lead to abnormal cell proliferation.

OTHER DISORDERS

The discussion of molecular pathology has centred, so far, on the haemoglobinopathies and immunoglobulin diversity, because of the general principles which their study has revealed. The next few sections will explain the way in which molecular studies have advanced our understanding of several important systems or structures. Sometimes the proteins involved were already well studied biochemically but their physiological and developmental role was unclear. Often the impetus for the research was to understand human genetic disorders and the availability of such naturally occurring mutations was very helpful, even though the disorders themselves were sometimes very rare. Many of the most important disorders clinically have been shown to arise from previously unknown gene products. These will be discussed later, in Chapter 7.

Skeletal disorders

Skeletal disorders largely result from mutations of collagen genes. Type I collagen consists of two types of alpha chains, type I and type 2 encoded by quite separate genes on different chromosomes. Mutations in either gene give rise to brittle bone disease, osteogenesis imperfecta, which is inherited in a dominant manner. Over

100 mutations have been identified resulting in a wide range of severity. Mature collagen molecules consist of a triple helix and one chain with a mutation will normally associate with other good chains resulting in so-called protein suicide. Mutations in type II collagen and type X collagen lead to a wide range of clinically separate disorders. These are shown in Fig. 5.10 together with mutations which have been created in mice which suggest that similar mutations may be found in humans eventually. Mutations of other collagen genes have been implicated in a wide range of disorders (Table 5.5) including the basement membrane collagen type 4 in the kidney disease Alport syndrome and type 3 collagen in the Ehlers–Danlos disorders leading to stretchy skin or weak cardiac walls leading to susceptibility to aortic aneurysms.

Collagen chains undergo extensive modification after translation. Mutations in several of the enzymes involved in these alterations can also cause genetic disorders involving collagen. A good example is the autosomal recessive disorder dermal dyspraxis, or Ehlers–Danlos type VIIC, in which the collagen fibrils in skin are disordered as a result of the failure of the proteinase, procollagen N-proteinase, to cleave the amino terminal portion from the developing chains.

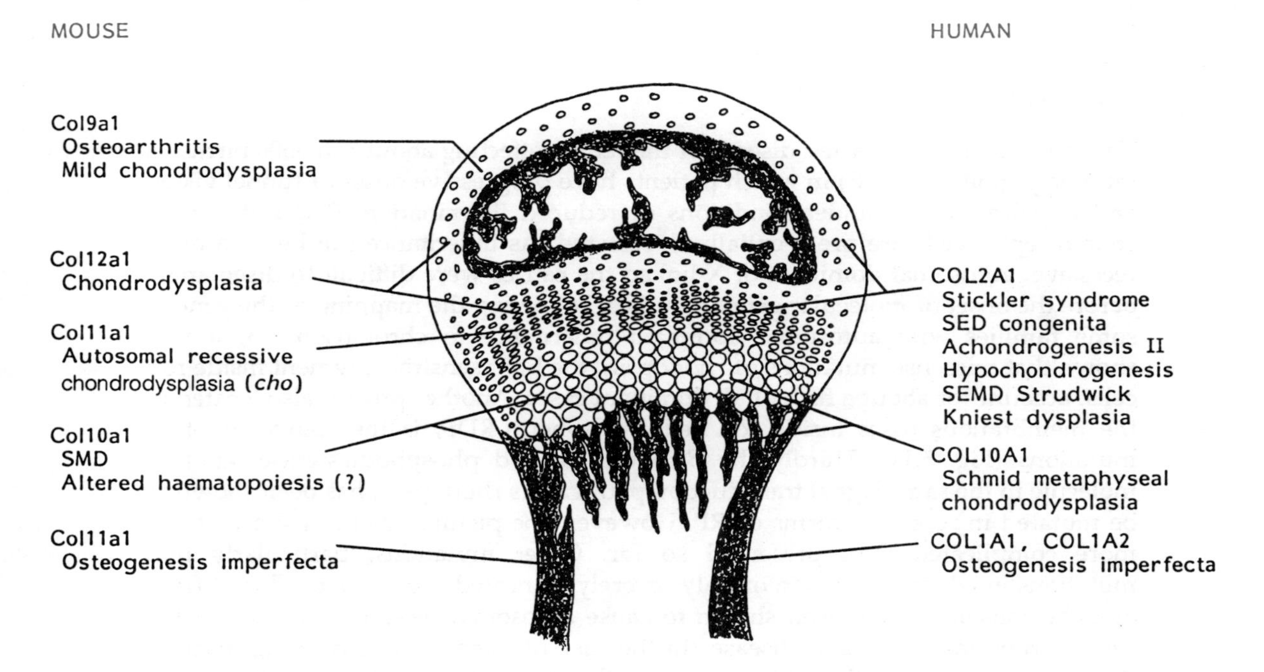

Figure 5.10 Developmental disorders of chondro-osseous tissue in mice and humans. Schematic representation of a developing long bone shows the epiphyseal and the metaphyseal zones. Collagens with mutations that lead to osteochondrodysplasias in mice and humans are listed. Lines indicate the region of tissue where the primary defects of these mutations are observed. (From Jacenko *et al.*, 1994, with permission.)

Table 5.5 Disease in which mutations of collagen genes or defective post-translational modifications have been shown to occur

Disease	Gene or enzyme
Osteogenesis imperfecta	COL1A1; COL1A2
Ehlers–Danlos syndrome type VIIA	COL1A1
Ehlers–Danlos syndrome type VIIB	COL1A2
Osteoporosis	COL1A1; COL1A2
Achondrogenesis	COL2A1
Hypochondrogenesis	COL2A1
Spondoepiphyseal dysplasia	COL2A1
Stickler syndrome	COL2A1
Kniest dysplasia	COL2A1
Osteoarthrosis	COL2A1
Ehlers–Danlos type IV	COL3A1
Aortic aneurysms	COL3A1
Alport syndrome	COL4A5
Diffuse leiomyomatosis	COL4A6
Dystrophic epidermolysis bullosa	COL7A1
Spondylometaphyseal dysplasia	COL10A1
Ehlers–Danlos type VI	Lysyl hydroxylase
Ehlers–Danlos type VIIC	Procollagen N-proteinase

Retinopathies

The most common inherited disease of the retina, affecting about 1 in 3500 births, is retinitis pigmentosa (RP) in which patients have progressive onset of tunnel vision and impaired vision under conditions of reduced illumination. Rod rather than cone receptor cells are preferentially devastated. As inheritance can be autosomal recessive, autosomal dominant or X-linked the causes were difficult to disentangle before the onset of molecular genetic studies. Following the mapping of the gene in some families with autosomal dominant inheritance to chromosome 3, it was shown that missense mutations of rhodopsin, the light sensitive pigment inside rod cells, account for about a fifth of all dominant cases. Another protein also located in the membranous discs inside rod cells, peripherin-RDS, is the source of other mutations (Fig. 5.11). Thirdly the β-subunit of rod phosphodiesterase, another molecule in the same signal transduction pathway as rhodopsin, has been shown to be mutated in recessive forms of RP. However, the picture has been shown to be more complicated than presented so far. Other mutations, particularly null mutations in which no protein or only severely truncated protein is produced from one chromosome, have been shown to cause autosomal recessive disease (in the case of rhodopsin) or mild disease (in the case of peripherin) involving macular degeneration in which the central region of the retina responsible for most daytime vision breaks down. Finally a conservative mutation of rhodopsin (Ala 292Glu) caused congenital stationary night blindness in a patient in which there was no degeneration of either rod or cone receptors. These results are summarized in Table 5.6.

These findings illustrate many interesting points including the variation in

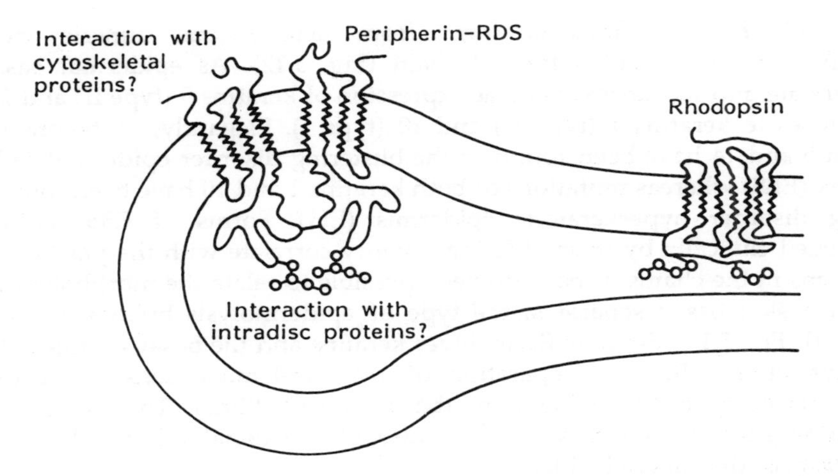

Figure 5.11 The end of a membranous disc. Rhodopsin is found in the central region. Peripheral RDS molecules, at the ends of the disc, are believed to be present as dimers. (From Wellcome Trust Annual Report, 1993, with permission.)

clinical presentation depending on mutation and the importance of the candidate gene approach.

Blistering and scaling diseases of skin

Keratin proteins are the major constituents of the epidermal protective structures such as hair and skin. Increasingly, we now understand how alterations in their structure affect skin stability and particularly how the level of skin involved reflects the developmental process of keratin expression. There are at least twenty keratin genes known but they fall into two classes, type I located in a cluster on chromosome 17q12–q21, and type II located in a separate cluster on chromosome

Table 5.6 Mutations in retinopathies

Gene	Disorder	Mutations
Rhodopsin	AD retinitis pigmentosa	Missense
	AR retinitis pigmentosa	Null
	Congenital stationary night blindness	Conservative amino acid change
Peripherin (or retinal degeneration slow)	AD retinitis pigmentosa	Missense
	Retinitis punctata albescens (Macular degeneration)	Heterozygous for null allele
	Vitelliform macular dystrophy (very mild)	Heterozygous for null allele
β-sub-unit rod phosphodiesterase	AR retinitis pigmentosa	Nonsense
	AD stationary night blindness	Heterozygous missense

AR = autosomal recessive.
AD = autosomal dominant.

12q11–q13. Keratin filaments form from heterologous dimeric molecules containing a type I and a type II chain (Fig. 5.12). As epidermal basal cells differentiate into suprabasal cells the expression of keratins 5 (type II) and 14 (type I) switches to keratins 1 (type II) and 10 (type I). Elegantly, mutations of both keratin 5 and 14 have been found in the blistering disorder epidermolyis bullosa simplex (EBS) whereas mutations of both keratins 1 and 10 have been found in the scaling disorder hyperkeratotic epidermis (EH). Forms of EBS which were recognized clinically by their differing severity correlate with the positions of the mutations in the chains. It has also been possible to relate the morphological level at which skin tissue separates, the type of epidermolysis bullosa and proteins involved (Fig. 5.12). Simplex EB involves keratins and the basal keratinocyte layer but dystrophic EB, with separation of the sub-lamina densa, is caused by mutations of type VII collagen in the anchoring fibrils. This suggests many candidate genes which may be the cause of junctional EB in which there is separation of the lamina lucida.

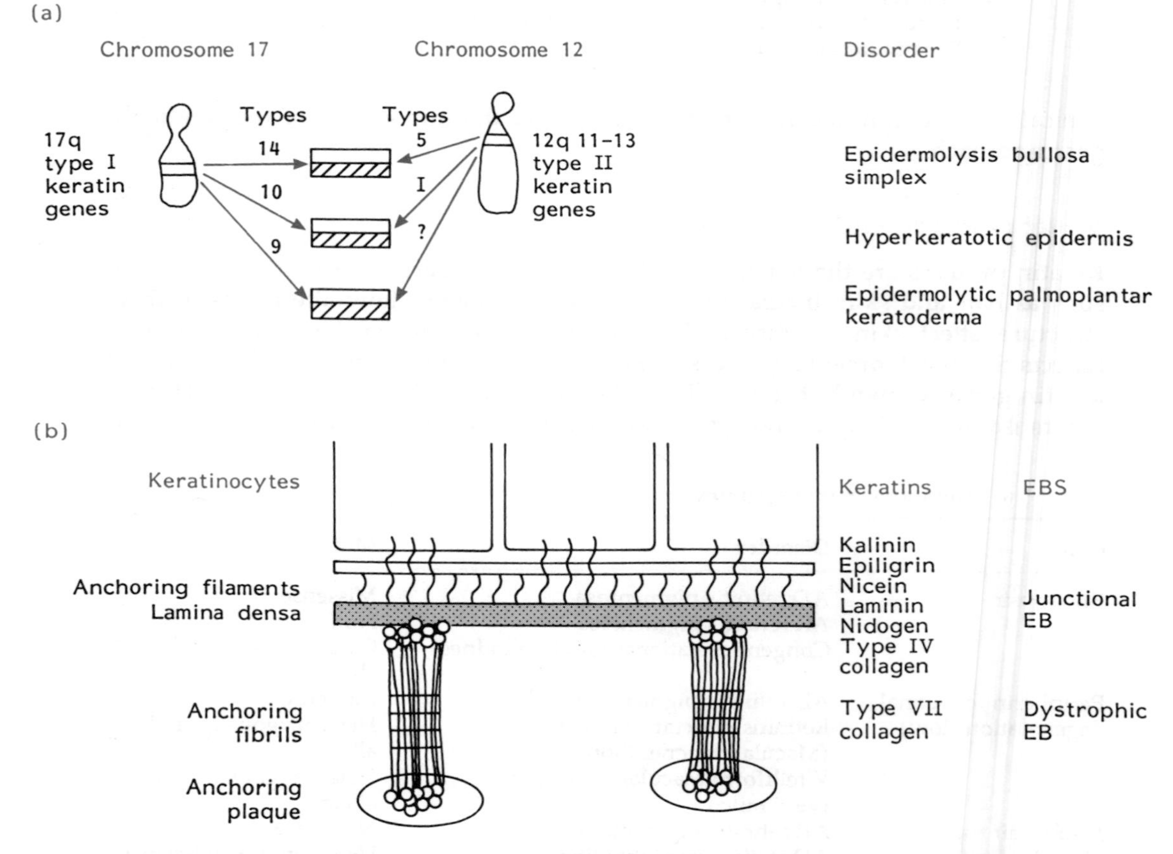

Figure 5.12 Keratin disorders. (a) Mapping of skin disorders to mutations in keratin genes. (b) Role of mutant genes in blistering disorders. (From Uitto and Christiano, 1992, with permission.)

Immunodeficiencies

A great deal of effort has been expended in recent years to understand the complex interactions of cytokines which result in the mature B and T lymphocytes necessary for a correctly functioning immune system. Complete understanding of these processes will allow insight into the development of leukaemias, allergies and infections as well as the development of therapeutic strategies.

The X chromosome contains genes for at least six immunodeficiencies (Table 5.7). Three of the genes which have been isolated in the last few years have improved our understanding of the development of T and B cells. Particularly interesting has been the study of hyper IgM syndrome in which patients produce B cells with IgM but fail to carry out the class switch to produce the full rage of antibodies. Instead of being a problem of the B cells, this is caused by a mutation of a T helper cell surface molecule (CD40 ligand) which is needed to interact with CD40 on the surface of B cells to stimulate the switch to IgA, E and G. Bruton's agammaglobulinaemia (XLA) is caused by a defect further back in B cell maturation. In the absence of a protein tyrosine kinase, btk, the correct cell signalling is missing to convert pre-B cells to B cells. The third important molecule to be identified was one of the sub-units of the IL2 receptor which was found to be mutated in patients with the X-linked form of severe combined immunodeficiency. These patients have a devastating, lethal, immunodeficiency which has now been explained by the finding that the mutated IL2γ chain also forms part of the receptor for the cytokines IL4 and IL7 and possibly others.

The recent identification of the genes involved in some inherited human immunodeficiencies has improved our understanding of the normal processes via these 'experiments of nature' and shown how the study of rare, or even very rare, diseases can teach us a great deal about basic biological processes.

Table 5.7 X-linked immunodeficiencies

Disorder	Symptoms	Gene	Function
Chronic granulomatous disease	Infections	Cytochrome oxidase *b-245*	Phagocyte killing
Properdin deficiency	Complement	Properdin factor B	Activation complement pathway
Wiskott–Aldrich	Eczema and infections	?	?
X-linked severe combined immunodeficiency	Lack of T cells	IL-2 receptor γ-sub-unit	Binding of cytokines
Bruton's agammaglobulinaemia	Lack of circulating B cells	*btk*	Protein tyrosine kinase
Hyper IgM	Elevated levels IgM Reduced levels IgG	CD40 ligand	T cell–B cell interactions
X-linked lymphoproliferative Duncan disease	Lymphocyte proliferation after Epstein–Barr infection	?	?

Disorders of myelin

It will scarcely come as a surprise to readers of the previous few sections that using the tools of molecular genetics the disorders caused by mutations in the major protein components of myelin are rapidly being identified (Fig 5.13). However, the myelin disorders have produced many surprises in another direction, namely in the range of types of mutation giving rise to very similar disorders and conversely the range of phenotypes caused by apparently rather similar mutations. The prime example is found in the hereditary motor and sensory neuropathies (HMSN), or Charcot–Marie–Tooth disease. HMSN type I is a peripheral neuropathy, inherited in a dominant manner, in which demyelination leads to a slowing of the nerve signal. A peripheral myelin structural protein (PMP-22) has been shown to cause

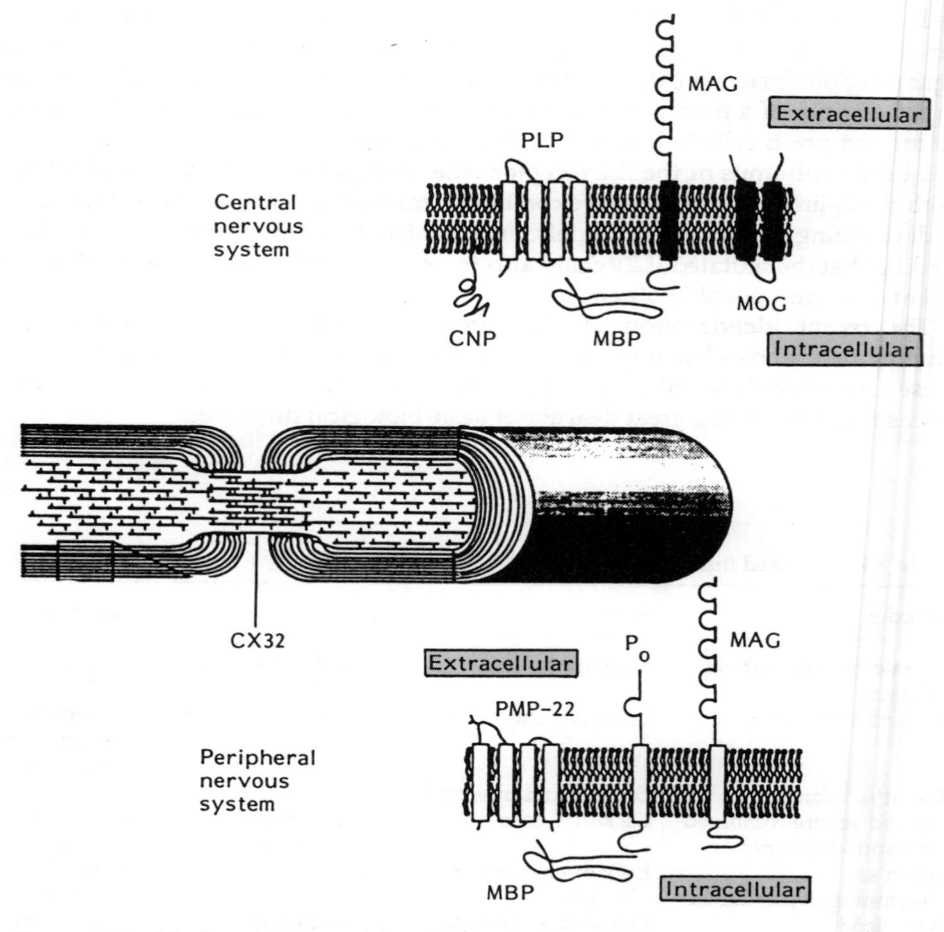

Figure 5.13 Myelin and its proteins. PLP, proteolipid protein; PMP-22, peripheral myeline protein 22; CX 32, connexin 32; P_0, protein zero; MBP, myelin basic protein; MAG, myelin associated glycoprotein; MOG, myelin/oligodendrocyte glycoprotein; CNP, 2'-3' cyclic nucleotide 3'-phosphohydrolase. (From Aubourg, 1993; Suter, 1993, with permission.)

HMSNI, but surprisingly in the majority of cases it produces the disease because of the presence of three copies of the gene rather than the normal two. This arises because of a large DNA duplication on the short arm of chromosome 17 which includes the PMP-22 gene (see page 87). In a minority of cases, point mutations which inactivate the gene cause the disease, but a complete deletion of the gene does not result in HMSNI but a much milder neuropathy in which temporary muscle weakness occurs after strain (hereditary liability to pressure palsies). The rules of the genetic code are neatly illustrated in a patient in which a 2 base pair deletion occurs early in the gene. As the genetic code is read in triplets this results in a very early frameshift (and presumably therefore, negligible protein) and also cause hereditary liability to pressure palsies (HLPP). Most confusingly, other point mutations give rise to a much more severe form of motor and sensory neuropathy, type III or Dejerine–Sottas. The relationship of these mutations to the function of PMP-22 will provide an interesting study for the next few years, but may be assisted by the finding that the most abundant peripheral myelin protein P zero (P0) can be mutated to give both type I and type III HMSN. A further HMSN, from its pattern of inheritance, is caused by a gene on the X chromosome. This has been shown to be caused by mutations in a gap junction protein connexin 32, localized to the nodes of Ranvier along the axon. Gap junctions form channels between cells, allowing ions and small molecules to transfer.

Remarkable similarities occur in a lethal dismyelinating disorder of the central nervous system, Pelizaeus–Merzbacher disease (PMD). This has an X-linked pattern of inheritance and arises from mutations of a gene localized on the X chromosome, proteolipidprotein (PLP), which is a major structural protein of CNS myelin. Missense mutations, duplication and deletion of PLP have all been found in affected boys. There is an alternatively spliced (see page 66) form of PLP, called D20, formed during development in which a splice site within exon 3 is activated and the latter section of exon 3(3b) forms part of the intron. A missense mutation within 3b which will affect PLP but produce normal D20 has been found in a boy with spastic paraplegia. These results are summarized in Table 5.8.

Genes involved in development

Around a decade ago, groups of genes were discovered in *Drosophila melanogaster*, a type of fruit fly, which are regulators of embryonic development and are responsible for determining the types of structures the various segments of the embryo will form (i.e. antennae in the head and legs in the thorax). The genes share a common motif, a 183 bp homeobox which encodes a 61 amino acid homeodomain responsible for DNA binding. They have been very highly conserved during evolution, both in structure and organization and are found in both man and mouse. In humans there are four clusters containing 38 genes and their pattern of expression along each cluster mirrors the sites of embryonic expression. No mutations causing birth defects have yet been found in the human homologues, although a further hox-related gene, *msx*-2, which shares some sequence similarity but which does not form part of one of the major clusters, has been found to be mutated in a family with a form of craniosynostosis or premature fusion of the cranial sutures. Another important family of genes was discovered in

Table 5.8 Mutations in hereditary motor and sensory neuropathies (HMSN) and Pelizaeus–Merzbacher disease (PMD)

Gene	Mutation	Disorder
Peripheral myelin		
PMP-22	Duplication	HMSNI
	Missense	HMSNI
	Splice site	HMSNI
	Deletion	HLPP
	Missense	HMSNIII
P_0	Missense	HMSNI
	Missense	HMSNIII
Cx32	Missense	HMSNX
	Frameshift	HMSNX
CNS myelin		
PLP	Missense	PMD
	Splice site	PMD
	Deletion	PMD
	Duplication	PMD
	Isoform	Spastic paraplegia

Drosophila and humans with a 128 amino acid conserved stretch. These were called paired box, or *pax*, genes after the original *Drosophila* gene in which the motif was found. Mutations in *pax*-3 result in Waardenburg syndrome, which is a hereditary disease leading to sensorineural deafness, premature greying and apparent lateral displacement of the eyes although translocations of the region of chromosome 2 containing *pax*-3 to chromosome 13 result in rhabdomyosarcomas, *pax*-6 has been implicated in aniridia, lack of formation of an iris, in humans. It remains to match up mutation of the other *pax* genes with recognized human malformation syndromes (Table 5.9). Fibroblast growth factor receptor genes (FGFR), a family of four related genes, are important for the correct formation of the skeleton and skull. A mutation in FGFR3 causes achondroplasia, the most common form of genetic dwarfism, and mutations in FGFR1 and 2 cause forms of craniosynostosis (or premature fusion of the skull futures).

Table 5.9 Genes involved in embryo development

Gene	Chromosome	Syndrome	
msx-2	5q	Human:	Craniosynostosis
pax-3	2q	Human:	Waardenburg rhabdomyosarcoma
		Mouse:	Splotch
pax-6	11p	Human:	Aniridia
		Mouse:	Small eye
FGFR3	4	Human:	Achondroplasia
FGFR2	10	Human:	Craniosynostosis (Crouzon) (Apert)
FGFR1	8	Human:	Craniosynostosis (Pfeiffer)

Imprinting

All the disorders discussed so far have assumed that an equal contribution will be made by a gene whichever parent it is inherited from. Although usually the case, this is not always so and we now know of examples where the gene is silenced in one parent, or imprinted. A normal individual will only form if there is a genetic contribution from both mother and father. The best studied examples to date are the severe childhood developmental disorders, Angelman and Prader–Willi syndromes. Both of these disorders are associated with a small cytogenetic deletion of chromosome band 15q11–13 which can sometimes be seen down the microscope in good preparations. The piece missing is the same in the two disorders and the only difference is that in Prader–Willi syndrome the deletion is on the chromosome inherited from the father and in Angelman syndrome on the chromosome inherited from the mother. In a further departure from the genetics described by Mendel some cases of Prader–Willi arise because the child has inherited two copies of chromosome 15 from their mother and none from their father. This is called uniparental disomy, i.e. two chromosomes from one parent, and occurs very rarely when as a result of an incorrect meiosis an embryo is formed initially with two maternal chromosomes and the usual single paternal but only survives because one of the chromosomes is lost in early embryogenesis. The net result is the same as a chromosomal deletion of the paternal chromosome, in both cases the child lacks a paternal contribution of gene(s) encoded in that region. Conversely, Angelman syndrome arises when the child inherits two copies of the paternal chromosome 15. Clearly there are two regions close together on chromosome 15 and imprinted in opposite directions. So far it has been established that some bases are differentially methylated depending on their parental origin but the genes affected, details of the mechanism and indeed the reason why the imprinting is necessary remain to be elucidated.

SUMMARY

- Recombinant DNA studies are rapidly revealing details of the molecular pathology of many unifactorial diseases. Initially most was learned from the study of the thalassaemias and they provide examples of most phenomena seen later in other disorders. Comparable defects have been detected in inborn errors of metabolism, developmental anomalies and structural defects.
- A genetic disorder may result not only from a point mutation within the coding (exon) part of the responsible gene but also by a variety of other mechanisms. These include point mutations of promotor sequences, stop codons and splicing sites; frameshifts; gene deletions; and gene fusions as a result of unequal crossing-over between adjacent gene loci.
- Different mutations in the same gene can have very different effects and, in several cases, disorders which were regarded as distinct entities from a clinical standpoint are now known to be allelic variants of the same gene.

- In some cases the opposite is true. Disorders which are clinically indistinguishable have been shown to be caused by completely different genes.
- The study of antibody formation illustrates yet another way in which genetic diversity may be produced, in this case through random, somatic recombination of genetic elements.
- Somatic recombination of immunoglobulin and T-cell receptor genes can sometimes accidentally involve other genes and, if the gene in question is a potential oncogene, this is an important cause of leukaemias and lymphomas.

REFERENCES AND FURTHER READING

Textbooks and review articles

Ballabio A. The rise and fall of positional cloning? *Nature genetics* 1993; **3**: 277–279

Emery AEH, Mueller R. *Elements of Medical Genetics*, 8th edn. Edinburgh and London: Churchill Livingstone, 1992

Epstein EH. Molecular genetics. Epidermolysis bullosa. *Science* 1992; **256**: 799–804

Francke U, Furthmayer H. Marfan's syndrome and other disorders of fibrillin. *Lancet* 1994; **330**: 1384–1385

Higgs DR, Weatherall DJ (eds) *Baillière's Clinical Haematology: The Haemoglobinopathies*, Vol 6:1. London: Baillière Tindall, 1993

Humphries P, Kenna P, Farrar GJ. On the molecular genetics of retinitis pigmentosa. *Science* 1992; **256**: 804–808

Leder P. The genetics of antibody diversity. *Sci Amer* 1982; **246**(5): 72–83

McKusick VA. *Mendelian Inheritance in Man*, 10th edn. London and Baltimore: Johns Hopkins Press, 1992

Weatherall DJ. *The New Genetics and Clinical Practice*, 3rd edn. London: Oxford University Press, 1993

Wellcome Report 1993

Research publications

Aubourg The leukodystophies: a window to myelin. *Nature Genetics* 1993; **5**: 105–106

Choo KH, Gould KG, Rees DJG, Brownlee GG. Molecular cloning of the gene for human anti-haemophilic factor 1X. *Nature* 1982; **299**: 178–180

Chu M-L, Williams CJ, Pepe G, Hirsch JL, Prockop DJ, Ramirez F. Internal deletion in a collagen gene in a perinatal lethal form of osteogenesis imperfecta. *Nature* 1983; **304**: 78–80

Giannelli F, Choo KH, Rees DJG, Boyd Y, Rizza CR, Brownlee GG. Gene deletion in patients with haemophilia B and anti-factor IX antibodies. *Nature* 1983; **303**: 181–182

Jacenko O, Olsen BR, Wardman ML. Of mice and men: heritable skeletal disorders. *Am J Hum Genet* 1994; **54**: 163–168

Robson KJH, Chandra T, MacGillivray RTA, Woo SLC. Polysome immunoprecipitation of phenylalanine hydroxylase mRNA from rat liver and cloning of its cDNA. *Proc Natl Acad Sci USA* 1982; **79**: 4701–4705

Suter U, Welcher AA, Snipes GJ. Progress in the molecular understanding of hereditary peripheral neuropathy reveals new insights into the biology of the peripheral nervous system. *TINS* 1993; **16**: 50–55

Uitto J, Christiano A. Molecular genetics of the cutaneous basement nerutorane zone. *J Clin Invest* 1992; **90**: 687–692

Chapter 6
Molecular pathology of some common diseases

It is useful to consider all human disease as being on a spectrum. At one end are those diseases which are genetically determined, and include the chromosomal and unifactorial disorders. At the other end are the environmentally determined conditions, which include nutritional deficiencies and infectious diseases. Between these extremes are many common disorders in which both environmental and genetic factors are involved in aetiology. It is believed that each of the disorders in this latter group are caused by many genes which render the individual susceptible to particular environmental factors, so-called **multifactorial** inheritance. Unlike unifactorial disorders, multifactorial disorders are common, their genetics are complex, the risks to relatives are usually low (less than one in twenty) and cannot be predicted on genetic theory but have to be determined empirically for each disorder. These disorders fall roughly into four categories: so-called 'diseases of modern society' (diabetes mellitus, atherosclerosis and coronary artery disease, hypertension and peptic ulcer), the commoner congenital malformations (such as spina bifida and anencephaly), certain psychiatric disorders (such as schizophrenia), and probably the commoner forms of cancer. In some of these disorders (such as schizophrenia and coronary artery disease) genetic factors seem to be more important in aetiology than in others (such as congenital heart disease and lung cancer). However, there are those who feel that the concept of multifactorial inheritance is losing some of its relevance because increasingly research is revealing considerable **genetic heterogeneity** in many of these disorders. By genetic heterogeneity is meant that a particular clinical picture may be produced either by genes at different loci or by different alleles at the same locus. Nowhere is this better illustrated than in the case of diabetes mellitus. It is therefore perhaps better to define these common disorders no more precisely than that they are often familial and there is an inherent susceptibility to a particular environmental agent, though in many cases this has not yet been defined.

In these disorders, recombinant DNA technology has much to offer but apart from cancer studies, which have benefited greatly, progress in other areas has been less dramatic and it has often proved very difficult to interpret the effects observed. This is partly because in almost all other common disorders the genes determining susceptibility have not yet been identified. Gradually some progress has been made, an outstanding example being Alzheimer's disease. A combination of approaches have been used including the rare example of families in which the disease follows a simple pattern of inheritance, cloning and analysis of candidate genes in pathways and the association of particular allelic forms of a candidate gene with the disease state more often than expected at random. A combination of approaches is needed because more than one gene is involved. So far genes causing this condition have been identified on chromosomes 21 and 14 in early onset cases and 19 in later onset cases. The contribution which recombinant DNA technology can make is in diagnosis, classification, molecular epidemiology and pathogenesis. A few examples will serve to illustrate some of these points, and it would seem worth while to begin with diabetes mellitus.

DIABETES MELLITUS

The term diabetes mellitus refers to a group of disorders in which the level of sugar in the blood and/or urine is abnormally high. Over sixty different unifactorial disorders are associated with glucose intolerance and in some cases with frank diabetes. Though many of these disorders are individually rare, this nevertheless indicates that there are many different loci involved in controlling the level of blood sugar. Which, if any, of these genes are involved in the pathogenesis of the common forms of diabetes is not known. Even the latter is genetically heterogeneous for **at least** two quite distinct forms are now recognized: insulin-dependent diabetes mellitus (IDDM) with juvenile onset and the much commoner non-insulin-dependent diabetes mellitus (NIDDM) with mature onset. Whereas individuals with IDDM require regular injections of insulin throughout their lives, those with NIDDM can often be controlled by dietary restrictions alone. IDDM is a more serious disease and is associated with specific HLA antigens, pancreatic islet cell antibodies often being present—features not found in NIDDM. Both forms show increased familial incidence concordant for the same disease, but apart from some rare forms of NIDDM the inheritance of these disorders is not clear. It also seems likely that even these two forms are not homogeneous.

Because IDDM is a far more serious disease it has understandably generated more interest from a research point of view, but the cause still remains obscure. One possibility, for which there is a little evidence, is that it may be the result of a viral (perhaps Coxsackie B) infection which results in pancreatic damage in those who are genetically susceptible and who then mount an autoimmune reaction to pancreatic islet cells which causes further damage. However, one of the main obstacles to research in this disease, and incidentally to its prevention one day, is to know which individuals are genetically predisposed to developing the disease, and here HLA typing may be helpful.

The HLA region on chromosome 6 (6p21–p23) consists of four different loci (A, B, C, D0), each with many alleles and since these loci are very closely linked they tend to be inherited en bloc (Fig. 6.1). The term **haplotype** is used to describe any group of closely linked alleles which are inherited together as a unit. In the case of the HLA system, a haplotype is the particular four HLA genes carried on each of an individual's two chromosomes 6. Affected sibs usually share the same HLA haplotype and an unaffected individual who shares a particular haplotype with his affected sib has an increased risk of developing disease. In IDDM about half the inherited predisposition is dictated by genes within the HLA region and these

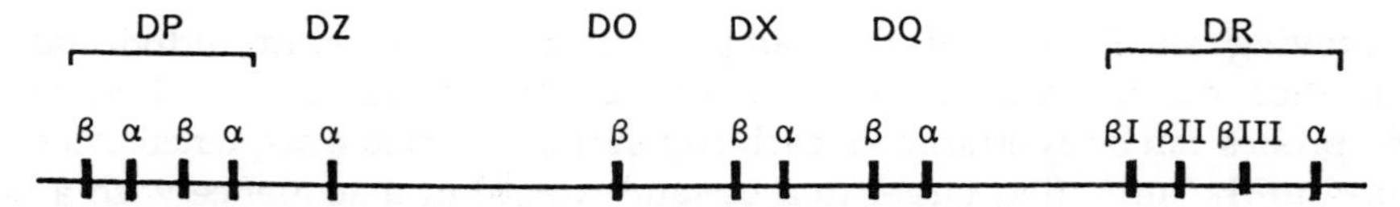

Figure 6.1 Map of the class II loci of the human A–D region on the short arm of chromosome 6. (From Todd *et al.*, 1987, with permission.)

genes can decrease the risk (protect against development of diabetes) as well as increase it. In particular the DQβ chain of the HLA region seems to be important in specifying the autoimmune response against the insulin producing islet cells. Todd and colleagues sequenced four major polymorphic HLA class II gene products in three patients with IDDM. They found that although there was no unique sequence in IDDM patients, amino acid 57 of DQβ correlated with susceptibility to IDDM. Interestingly a mouse model for diabetes, the NOD mouse, has a *ser* 57 mutation also. Following this they looked at the DQβ sequence in a large number of individuals with and without IDDM. They used a technique called **allele specific oligonucleotide** hybridization of **ASO** (Fig. 6.2). This involves amplifying the exon from the gene using the PCR so that there is plenty of target sequence to probe

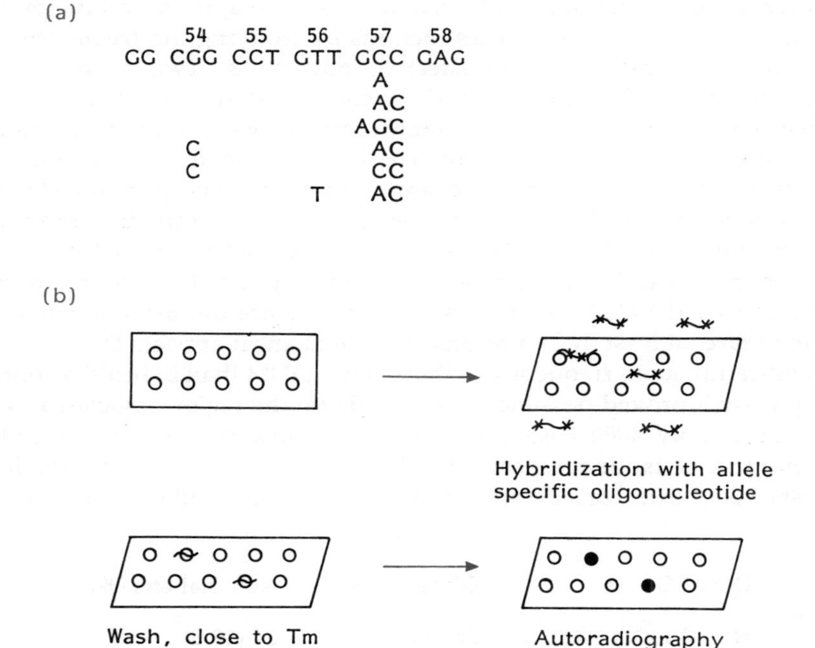

Figure 6.2 (a) Range of oligonucleotides from exon 1 of HLA-DQβ used to detect different alleles. (b) Exon 1 DQβ PCR products. XNX = Radioactively labelled oligonucleotide; Tm = melting temperature.

and applying small dots of the samples from the different individuals onto hybridization membranes in rows and columns (**dot blotting**). A whole range of specific probes can be synthesized, each containing a variable sequence. All of them will bind to the amplified target to a certain extent but a sequence with a perfect match will be more stable and have a higher melting temperature than a probe with one or two mismatched base pairs. A probe of about 17 nucleotides (hence the term oligonucleotide) gives the best results because it will bind specifically but be short enough that the melting temperatures will vary sufficiently to distinguish perfect matches from partial mismatches. The probes are radioactively labelled, hybridized with the filter and then the filter washed paying great attention to the temperature of the wash solution. Altogether Todd and colleagues used eight specific oligonucleotide probes corresponding to known HLA DQβ sequences. The most striking result was that 96% of the diabetic probands had a genotype in which neither of their two copies has the amino acid Asp at position 57. In fact, no diabetic patient had an aspartic acid at this position on both chromosomes. In contrast, 42% of the non-diabetic controls had *asp* 57 on both chromosomes (Fig. 6.3). In other words, the presence of an HLA DQβ chain with *asp* 57 protects against developing IDDM.

The human insulin gene is located at 11p15, about 14 cM from the β-globin gene, and has been isolated, cloned and sequenced. It is 1430 base pairs in length with two introns. About 360 base pairs upstream from the starting point of transcription is a region with a very high frequency of DNA length polymorphisms, containing variable numbers of a 14 base pair repeat, and therefore referred to as a **hypervariable region**. A number of investigators have approached the problem of determining susceptibility to diabetes mellitus by studying the frequency of DNA polymorphisms in this region in affected individuals. Two such studies will illustrate this point. Peripheral blood leucocyte DNA was digested with an appropriate enzyme and then the resultant fragments were hybridized to a labelled insulin gene probe on a Southern blot. The DNA fragments so identified contain not only the insulin gene itself but also several thousand base pairs on either side of the gene. Depending on the fragment sizes generated by restriction enzymes, Bell and Karam (1983) found three classes of alleles defined by their different lengths due to the numbers and arrangements of tandem repeats at the polymorphic locus (Fig. 6.4). Class 1 (L) alleles and class 3 (U) alleles are the most frequent, class 2 alleles being rare, at least in Caucasians. In Caucasians it appears that class 1 alleles occur significantly more frequently in those with IDDM than in healthy controls.

Subsequently it proved possible to narrow down the region associated to a much shorter area of only 4000 base pairs around the insulin gene and including the variable number of tandem repeats (VNTR). This time not only was the length of the repeats taken into consideration but a haplotype of other base changes was

DQ ASO	Diabetic (%)	Non-diabetic (%)
Non-Asp/Non-Asp	26 (96)	24 (19.5)
Non-Asp/Asp	1 (4)	57 (46)
Asp/Asp	0 (0)	42 (34)

Figure 6.3 DQβ variants in patients with IDDM.

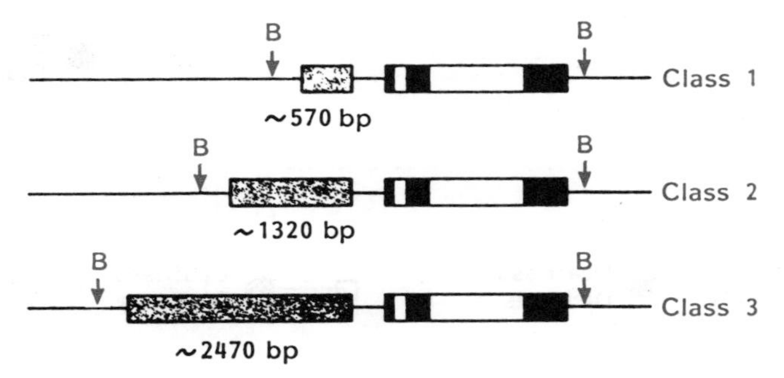

Figure 6.4 Diagrammatic representation of the polymorphic locus flanking the insulin gene which defines three classes of alleles. B = *Bgl*I site. (From Bell and Karam, 1983, with permission.)

constructed. The critical region does not include the nearby insulin growth factor receptor. However, it still seems that the polymorphic loci in the 5' flanking region of the insulin gene are markers for another gene(s) which influences susceptibility to diabetes rather than the insulin gene itself. Not only is there no evidence at all to suggest that insulin *per se* causes pancreatic damage but there is also no evidence that the number of repeats in the VNTR affects either the structure or amount of insulin produced.

A few specific gene changes have been found in NIDDM, or at least in some clinically recognized subtypes, involving different stages in the pathway of insulin activity and indeed occasionally in the structure of insulin itself. These changes are summarized in Table 6.1. One rather interesting variant results from a mutation altering the cleavage site of C peptide so that proinsulin accumulates rather than the mature insulin product.

It is probable that other genes for susceptibility to diabetes reside elsewhere in the genome. The problem is to know where to look. Studies using mice models allow very large pedigrees to be constructed which will allow the identification of modifying genes by linkage studies.

Further complexities have been added to the story by the observation that diabetes together with deafness was identified with a strong hereditary component in a small number of families. A typical pedigree is shown in Fig. 6.5. The disorder is inherited in a dominant manner, but close examination shows that although both sexes can be affected, the disorder is only ever passed on by a mother. Of course,

Table 6.1 Single gene changes in NIDDM

Glucokinase	MODY (mature onset diabetes of youth)
Insulin receptor	Insulin resistance
Insulin	NIDDM
Preproinsulin	Hyperproinsulinaemia

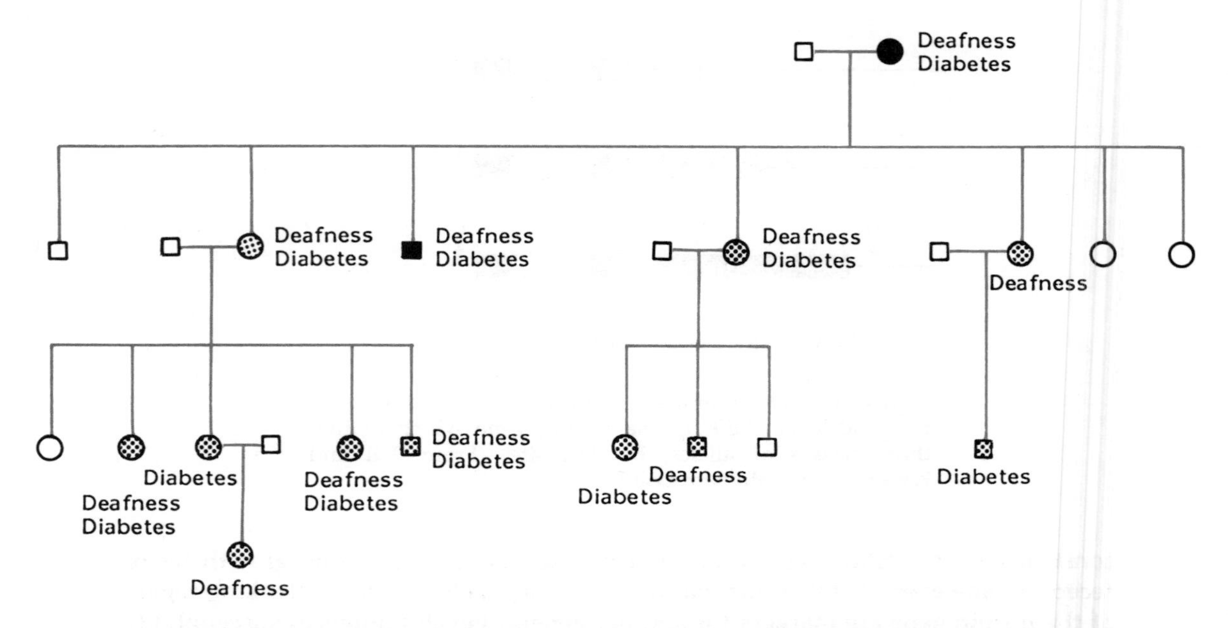

Figure 6.5 Family with individuals showing diabetes and/or deafness and a pattern of inheritance of mitochondrial disorders (i.e. maternal inheritance only).

none of the autosomes or the X or Y chromosome fit this pattern of inheritance. However, mitochondria are passed on through the mother's ova and any mutations carried on the mitochondrial genome will be expected to show maternal inheritance alone. They will also be expected to show quite a variation in the range of disease severity as there are many mitochondria in each cell and the proportion carrying the mutation may also vary (heteroplasmy).

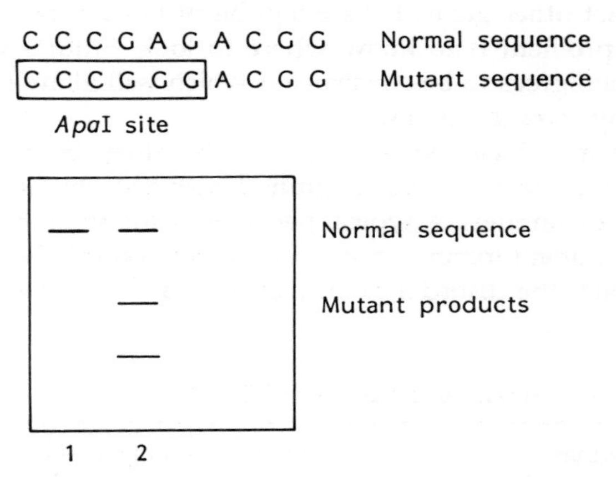

Figure 6.6 Detection of mutant mitochondria in patients with diabetes and deafness. 1, Uncut fragment. 2, Subject heteroplasmic for A → G mutation.

Mutations of the mitochondrial genome, particularly one involving the substitution of guanine for adenine at position 3243 of leucine tRNA (one of the 22 tRNAs encoded on mitochondrial DNA) have been found in several families with diabetes (both IDDM and NIDDM) associated with deafness. This cause of diabetes may be more common than previously expected as this mutation alone was found in 5/155 individuals with a family history studied. As this mutation occurs at a restriction enzyme recognition site for the enzyme *Apa*I, screening can be carried out rapidly by PCR amplification of a short region around the mutant site, digestion of the products with *Apa*I and analysis of the products by electrophoresis through an agarose gel (Fig. 6.6). The basic methods described for studying diabetes, i.e. association studies, candidate genes and rare familial forms provide the basis for study of most of the other common disorders to be described here.

ATHEROSCLEROSIS AND CORONARY ARTERY DISEASE

Atherosclerosis, or hardening of the arteries, is due to the formation of lipid deposits in the lining of the arteries, which predisposes to local thrombosis and which later coalesce, become fibrotic and eventually calcify. The consequent narrowing of involved arteries results in ischaemia of the tissues they supply with the development of coronary artery disease and cerebrovascular disease. Atherosclerosis is part of the natural ageing process with symptoms usually developing in middle or old age. However, like diabetes mellitus this is a genetically heterogeneous disorder and some patients develop coronary artery disease as early as the second decade of life. Several rare unifactorial disorders are associated with premature vascular disease, the commonest among these being autosomal dominant hypercholesterolaemia. It has been estimated that about one person in five hundred in the general population, and about one in twenty patients with ischaemic heart disease, are heterozygous for this disease, which is now known to be due to defective low-density lipoprotein receptors on cell surfaces with the result that the uptake of cholesterol is reduced and therefore plasma levels are elevated.

However, the majority of cases of coronary artery disease are not inherited in any simple manner, though there is clearly a familial component since the risk to first-degree relatives of an affected individual is on average six times greater than the population incidence, the latter being about 15 per 1000 in males with onset before the age of 55, and 10 per 1000 in females with onset before 65. There is general agreement that several different genes are probably involved in rendering an individual susceptible to atherosclerosis and coronary artery disease, additional important environmental factors being cigarette smoking and overeating. As in the case of diabetes mellitus, however, the problem is to know which gene(s) is of primary importance in determining susceptibility.

The genetic control of lipoproteins is clearly important since there is an association with raised plasma levels of low-density lipoproteins (LDL) and reduced plasma levels of high-density lipoproteins (HDL). It seems that while LDL may predispose the individual to atherosclerosis, HDL affords protection against the disease. Conversely low plasma HDL contributes to the development of premature atherosclerosis. Apolipoprotein AI (apoAI) is the major protein constituent of HDL and deficiencies of this particular protein, which is synthesized mainly in the liver, result in low plasma levels of HDL. There is evidence that depressed levels of apoAI provides the best correlation with coronary artery atheroma. This would therefore seem to be an excellent candidate for molecular studies in this disease. The normal gene for apoAI has now been isolated, cloned and characterized, the coding region of the mature apoAI mRNA consisting of 804 nucleotides. It has been found using apoAI probes that, at least in some individuals, premature atherosclerosis is associated with a variant apoAI gene with an insertion in the coding region. This in some way then interferes with expression of the gene and results in abnormal HDL levels. The major protein component of the LDL particle is apoB-100. Several restriction fragment length polymorphisms in and around the apoB gene have been identified and population studies in several different countries have shown that variation at an *Xba*I site within the gene is associated with differences in serum cholesterol levels. The presence of the site (*Xba*+) is associated with raised levels. The contribution to the variation in serum cholesterol levels from this source is 3–8% and it is, therefore, one of several such effects. The molecular basis of the effect is unknown as the base change that creates the *Xba*I site does not alter an amino acid as it is in the third (wobble) position of the codon for threonine 2488 and so cannot be the direct cause of the effects seen.

These examples serve to illustrate the molecular approach to the disease, but there are many other possible candidate genes (e.g. other apoproteins, fibrinogens, LDL receptors, etc.) which are all currently being studied. Class 3(U) alleles at the polymorphic locus flanking the insulin gene have been found to be associated with extensive atherosclerosis which is in addition to the conventional risk factors of obesity, hypertension, blood glucose and blood lipids. It may well be that an individual's susceptibility to atherosclerosis can best be determined by using several different DNA markers simultaneously. The results, if carried out early in an individual's life, might hopefully then influence his or her choice of future life-style and habits.

This general approach of association studies, discussed already for the insulin gene, apoAI and apoB, to reveal genes contributing towards complex disorders has

Table 6.2 Associations between gene variants and common disorders

Gene	Disorder
Tissue growth factor α	Cleft lip +/− cleft palate
Tissue growth factor α	Cleft palate only
Vitamin D receptor	Osteoporosis
Angiotensin-converting enzyme	Longevity
Angiotensinogen	Hypertension
Angiotensinogen	Pre-eclampsia
Apolipoprotein E4	Ischaemic heart disease (increased risk)
Apolipoprotein E4	Alzheimer's disease
Apolipoprotein E2	Ischaemic heart disease (lowered risk)
Apolipoprotein E2	Longevity
β-subunit IgE receptor	Atopy

been used widely. A summary is given in Table 6.2. Although a number of associations have been found in no case has the clear path from genetic variation to alteration in amount or sequence of product to cause of disorder been followed to the end.

NEUROGENETIC AND NEUROPSYCHIATRIC DISORDERS

The psychiatric disorders, as well as the commoner neurological disorders of unknown aetiology, pose difficult problems for the clinician, not least because of uncertainties over diagnosis and classification. The establishment of a precise diagnosis is essential for a number of reasons. Clinically similar but basically different disorders may well have different prognoses, respond differently to any possible treatment (and therefore confound assessment of the effects of treatment), and if there is any significant genetic component to aetiology the risks to relatives may be very different. Many exciting initial observations of linkage between various psychiatric disorders have collapsed after other groups have failed to reproduce them. This is probably because of problems of diagnosis in the initial studies but may, of course, suggest that the model assuming a single gene inherited in a Mendelian fashion is wrong.

A good example of such heterogeneity, but one where there is very exciting progress, is Alzheimer's disease. 5–10% of the population over 65 are affected with Alzheimer's disease. It starts with loss of recent memory and attention and progresses relentlessly with increasing loss of judgement and reasoning until death occurs in a vegetative state. Formal genetic studies are difficult to carry out because of the late onset, meaning that individuals may die of other causes before developing the disease, and similarities to other dementias. Senile plaques with amyloid deposits and neurofibrillary tangles accumulate in the brain. A major component of the plaques is β-amyloid peptide and the first breakthrough in unravelling the genetic causes of Alzheimer's came when linkage was shown, in a small number of families with early onset, to a region of chromosome 21 containing the gene for β-amyloid peptide. Several mutations of this gene were found in families, including three different changes, Val–Ile, Val–Phe and Val–Gly at amino acid 717. A further, so far unidentified gene, has been mapped to chromosome 14. However, the vast majority of Alzheimer's cases are of late onset and do not come in these categories, although even if the β-amyloid product itself is not altered interacting factors may be.

Even with the problems outlined above, linkage analysis suggested that there could be a gene on chromosome 19 involved in late onset cases. The different approaches began to come together when biochemical studies showed that apolipoprotein E (apoE) is localized in the senile plaques and binds to synthetic β-amyloid. However, there are several slightly different forms, or polymorphic variants, of apoE, E3 with a cysteine at position 112, E4 with the Cys 112 changed to Arg and E2 with Arg 158 changed to Cys. Of these apoE4 binds fastest to β-amyloid. The gene for apoE is on chromosome 19. Naturally, association studies followed fast and it was established that (1) there was a highly significant association of allele apoE4 and late onset Alzheimer disease, (2) the age of onset decreases with increasing dosage and (3) the apoE4 dose is related to survival. A total of 80% of familial and 64% of sporadic cases have at least one E4 allele compared to 31% of control subjects. Mean age of onset was 84.3 years in subjects with no E4 allele, 75.5 in subjects with one E4 allele and 68.4 in subjects with two alleles. This identifies the first genetic risk factor for Alzheimer's disease and shows that many apparently sporadic cases have in fact a genetic basis.

A further example of this approach has been the study of motor neurone disease (amyotrophic lateral sclerosis or Lou Gehrig disease). This is characterized by onset usually in middle age of progressive muscle wasting and weakness which results from degeneration of the anterior horn cells of the spinal cord and bulbar motor nuclei with involvement of the corticospinal tracts. In general, most cases are not believed to be genetic but again the study of a few families with a strong genetic predisposition showed linkage to a gene on chromosome 21, subsequently shown by mutation analysis to be superoxide dismutase (SOD (SOD1)). The normal role of SOD is to remove superoxide free radicals which can otherwise cause damage to cells. The mutations found will all involve loss of SOD activity presumably leading to dangerous free-radical build up. Why the free radicals should damage so specifically particular nerve cells remains a mystery. The same mutations account for at least a proportion of isolated or sporadic cases though probably less than 5%.

Several human neurodegenerative disorders have been shown to have long incubation times but instead of being caused by conventional viruses are caused by infectious prion particles which are unique in containing no nucleic acid but an abnormal isoform of the prion protein. Prions are transmissible and cause Kuru, Creutzfeldt–Jacob (CJD) and Gerstmann–Straussler–Scheinker (GSS) diseases in humans and scrapie and bovine spongiform encephalopathy (BSE) in animals. Kuru was once a common form of death in women and children in New Guinea who acquired it through ritualistic cannibalism. Since this practice stopped it has almost disappeared. Great publicity has been attached to the risk from so called 'mad' cows infected with BSE. In Great Britain it appears to have spread through herds during the late eighties by feeding infected offal to cattle. Naturally fears arose that the infectious agent may be able to cross the species barrier into humans. It will be many years before the evidence for or against this hypothesis becomes available. The gene for the prion protein maps to chromosome 20 and point mutations have been established in genetic forms of CJD, GSS and fatal familial insomnia. Thus prions can cause both transmissible and genetic forms of the diseases of which about 10% of CJD cases are genetic. Mutation of a particular codon, 102, has been found in GSS cases from many parts of the world whereas in CJD variable copy numbers of an octapeptide within the coding region have been observed. For many years a high incidence of CJD among Jews of Libyan origin was put down to the consumption of lightly cooked sheep brains or eyeballs. However, at least some Libyan and Tunisian Jews have been found to have a mutation involving amino acid 200.

Returning to the problem of psychiatric or behavioural disorders, many of the human genes responsible for the various enzymes involved in metabolic pathways within the central nervous system have been cloned. The hope is that these cloned genes could be used in two ways. Firstly, to detect genes which may be different in some way in psychiatric disorders as an aid towards the diagnosis and classification of these disorders. Secondly, by studying the structure and transcription of these genes from affected individuals it could throw light on their molecular pathology, as has happened in the thalassaemias, for example. The problem, as always, is to know which of the many enzymes involved should be selected for detailed investigation or, if the molecular defect resides in a neuropeptide which has yet to be identified, where to start the positional cloning efforts. Two examples may be taken from the seemingly unpromising areas of aggressive behaviour and schizophrenia. Painstaking studies in the Netherlands revealed a family in which there was a quite extraordinary clustering of men who had at some time in their lives shown unprovoked violent behaviour ranging from rape of a sister to arson and attempts to run over their boss. The women in the family did not show this aberrant behaviour and collection of all the pedigree data showed a characteristic X-linked pattern of inheritance. Enough family members were available and willing to participate in the study for a linkage study to be carried out which showed that the inheritance could be explained by a gene coded on the short arm of the X chromosome just above the centromere, in the region where the genes for monoamine oxidase A (MAOA) are found. As monoamine oxidases are involved in the metabolism of neurotransmitters this gene is a likely candidate for the disorder in this family. This is of great importance, not because genetic changes of this gene are likely to be common in aggressive males, but because it suggests that variations in the enzyme levels of MAOA from completely non-genetic sources may dangerously affect behaviour. Because much is known of the behaviour of neurotransmitters and MAOs therapeutic strategies could readily be devised. In the example of schizophrenia, a unique family has been identified in Scotland in which many affected members (and very few unaffected members) carry a balanced chromosome translocation between chromosome 1 and 11. This at least suggests a starting place in the search for candidate genes but the technical difficulties of the approach should not be underestimated and it seems likely to be some time yet before a major gene for susceptibility in schizophrenia is cloned and characterized.

Both these and a few other common disorders in which a single contributing gene has been identified in at least a minority of families are presented in Table 6.3.

Table 6.3 Common disorders, examples of single gene contributions

Motor neurone disease	Superoxide dismutase
Alzheimer's	β-amyloid Apolipoprotein E4
Creutzfeldt–Jacob	Prion protein
Gerstmann–Straussler–Scheinker	Prion protein
Psychotic behaviour	Monoamine oxidase
Aortic aneurysm	Type 3 collagen
Hypertrophic cardiomyopathy	β-myosin α-tropomyosin Troponin T
Supravalvular aortic stenosis	Elastin
Thrombosis	Protein C

SEXUAL ORIENTATION

Perhaps the most surprising and controversial trait for which a major contributing gene has been proposed, and almost certainly that with the longest term ramifications, is sexual orientation and specifically male homosexuality. This is an area in which even allowing for prejudiced views it is particularly difficult to disentangle genetic from environmental and cultural factors and it will, therefore, repay studying the methodology used.

Dean Hamer and his colleagues initiated the study by taking family histories from 76 index patients who were self-acknowledged homosexual men over the age of 18. The first problem to be tackled was whether there was a suitable diagnostic criterion for homosexuality. They used a scale rating several features of sexuality including self-identification, attraction, fantasy and behaviour. The first encouraging feature was that subjects fell into two distinct groups, implying that there is a real behaviour phenotype, not just a continuous variation. As traits that are genetically influenced aggregate in families, the homosexual participants were asked to rate their male relatives (subsequently 69/69 relatives who had been rated homosexual verified this themselves), 13.5% of brothers were found to be homosexual against 2% in the general population. Even more significantly maternal uncles and the sons of maternal aunts also had significantly higher rates of homosexuality, which is exactly in accordance with an X-linked pattern of inheritance (Fig. 5.2). Although they share genetic material they have not been raised in a common environment. Linkage analysis provides the means to test for X-linked inheritance and to distinguish it from maternal effects (such as upbringing) or imprinting. If there is an X-linked gene then the brothers should share the same region of the mother's X over some part of it. If there is an environmental effect then brothers inheriting either X should be affected evenly. Forty sib-pairs were analysed for the inheritance of markers along the X chromosome and in a sub-set of 33 the tip of the long arm of the X chromosome, band Xq28, was shared between the brothers.

This study awaits confirmation. In the mean time one should have an open mind as to whether or not such complex behaviour could be the result of a single gene on the X chromosome, implying as it does, that homosexuality is part of normal genetic variation.

CANCER

A malignant tumour or neoplasm consists of an abnormal and uncontrolled proliferation of cells which invade surrounding tissues and later metastasize to other parts of the body. Malignant tumours can be divided into two groups: those which arise from epithelial tissues and are referred to as carcinomas and those which arise from non-epithelial tissues (bone, muscle, connective tissue) and are referred to as sarcomas. The term cancer is often used loosely for any form of neoplasm irrespective of its origin, and it is in this sense that it will be used here.

A great deal has been learned and written about cancer over the years, and it would be impossible to cover the subject in any depth. Our main concern will be with some of the new approaches to our understanding of cancer which have been made possible through recombinant DNA technology. Even here it will not be possible to go into any great detail but only deal with general principles and some of the main findings. In general, we now recognize two types of cancer-causing genes: proto-oncogenes which are normal genes which undergo structural alterations, for example by chemical carcinogenesis or chromosome translocation, and anti-oncogenes or tumour suppressor genes. These latter genes are proteins whose normal function is to suppress uncontrolled cell growth (i.e. cancer) and their loss or inactivation results in the normal controls being lost. Usually both copies must be lost for tumours to be developed (recessive action) and so this class accounts for most inherited cancers where one copy of the gene is inherited in a mutated state in the germ line and the second copy acquires a somatic mutation during life. But first it is necessary to consider the role of genetic factors in cancer.

Genetic and environmental factors in cancer

In discussing the role of genetic factors in cancer, it is useful to consider the matter under four headings: rare genetic forms of cancer, inherited common cancers, the so-called 'cancer family syndrome' and finally the common cancers.

There are several rare unifactorial disorders which are associated with neoplasia (Table 6.4). Here environmental factors are, on the whole, of less importance in the development of cancer than the individual's genetic predisposition which is paramount. Since the risks to relatives are high, the recognition of these disorders is essential for genetic counselling. In a few of these disorders the molecular basis has already been worked out in detail as, for example, in xeroderma pigmentosum which is due to defects in the repair of DNA damage caused by ultraviolet light. Certain neoplasms also occur with increased frequency in chromosomal disorders, the most noteworthy being acute leukaemia in Down's syndrome (trisomy-21), breast tumours in Klinefelter's syndrome (XXY) and uterine tumours in Turner's syndrome (XO).

The second group includes very rare families where a common cancer is inherited as an autosomal dominant trait, affected individuals with the same type of cancer occurring over several generations of a family. An example is ovarian cancer. We have studied a family in which the disease affected seven females over three generations, in two cases being inherited through males.

The third group comprises the so-called 'cancer family syndrome' which has been defined largely by Lynch and his colleagues in the United States. The term is used for cancer-prone families in which the predisposition appears to be inherited as an autosomal dominant trait, tumours develop in early life (usually before the age of 40) and there are often multiple primary tumours at different sites. The various affected family members may well not have the same tumour. The concept now seems to be widely accepted though the difficulty is to distinguish these apparent cancer-prone families from a familial aggregation of common cancers which could occur by chance.

Finally, and by far the largest group, are the common cancers which have a multifactorial basis – a genetic predisposition to an environmental carcinogen(s). It is in this group that public health measures, through the recognition of environmental carcinogens and their subsequent avoidance, are most likely to have

Table 6.4 Some unifactorial disorders associated with neoplasia

	Clinical features	Neoplasia
Autosomal dominant		
Basal cell naevus syndrome	Naevi, skeletal abnormalities, jaw cysts	Basal cell carcinoma
Diaphyseal aclasis	Multiple exostoses	Sarcoma
Multiple endocrine neoplasia (MEN) type I	Hypercalcaemia, hyperinsulinaemia	Parathyroid, pancreatic islet cell and pituitary hyperplasia/ adenomas
Multiple endocrine neoplasia (MEN) type IIa	Phaeochromocytoma, parathyroid hyperplasia	Thyroid medullary cell carcinoma
Multiple endocrine neoplasia (MEN) type IIb	Phaeochromocytoma, mucosal neuromas	Thyroid medullary cell carcinoma
Neurofibromatosis I	Café-au-lait spots, neurofibromas	Neurofibromas, CNS tumours
Neurofibromatosis II	—	Acoustic neuromas
Polyposis coli	Colonic polyps	Colonic carcinoma
Retinoblastoma (bilateral)	—	Retinoblastoma
Tuberous sclerosis (epiloia)	Mental retardation, epilepsy, adenoma sebaceum	Benign tumours of retina, heart and kidney
Tylosis	Thickening of palms and soles	Oesophageal carcinoma in some families
Von Hippel–Lindau syndrome	Cysts of pancreas, liver and kidney	Haemangioblastomas of spinal cord and cerebellum
Autosomal recessive		
Albinism	Absence of pigmentation	Skin carcinoma
Ataxia telangiectasia	Ataxia, telangiectases, sino-pulmonary infections	Lymphoreticuloses
Bloom's syndrome	Dwarfism, light sensitive rash	Leukaemia
Chediak–Higashi syndrome	Partial albinism, photophobia, leucopenia, infections	Lymphoma
Fanconi's anaemia	Pancytopenia, congenital malformations, skin pigmentation	Leukaemia
Xeroderma pigmentosum	Freckle-like hypersensitivity to light	Skin carcinoma
X-linked recessive		
Bruton's hypogammaglobulinaemia	Bacterial infections, absence of plasma cells and Ig	Lymphoreticuloses
Wiskott–Aldrich syndrome	Eczema, thrombocytopenia, infections, bloody diarrhoea	Lymphoreticuloses

the greatest effects. Nevertheless, the recognition of genetic factors even in common cancers has two important implications. Firstly, it might provide a means for identifying individuals who are at particular risk. Secondly, and perhaps more importantly, it might lead to the development of rational therapy based on molecular pathology which could well be more effective than current methods. However, though all cancers are essentially genetic at the molecular level in that there is loss of the normal control of cell division, there appears to be no major inherited component in common cancers.

The interplay between environmental and genetic factors is well illustrated in the case of lung cancer. The overwhelming importance of environmental factors in causation is the incontrovertible link with cigarette smoking and with occupational exposure to such substances as asbestos, chromium, arsenic and nickel. Genetic factors are of much less importance but are evidenced by family studies which indicate that the risk of developing lung cancer in a heavy smoker is increased even more if a close relative has had the disease. However, the results of such studies are often difficult to interpret because of the confounding effects of shared family environments. More direct evidence of the role of genetic factors comes from work on the complex enzyme system referred to as aryl hydrocarbon hydroxylase (AHH) which breaks down certain polycyclic hydrocarbons (constituents of tobacco smoke) to highly active carcinogens. The inducibility of AHH, which is under polygenic control, has been shown to be associated with the development of lung cancer in cigarette smokers.

Evidence of the role of genetic factors at the cellular level in common cancers has been reviewed (see end of chapter).

Viruses and cancer

Over the last few decades research into the cause of human cancer has centred mainly on epidemiological, biochemical and cytological studies, but the new developments from recombinant DNA work have largely stemmed from research on tumour viruses. The seminal discovery in this field was made as long ago as 1910 when Peyton Rous of the Rockefeller Institute for Medical Research showed that a cell-free filtrate from a chicken sarcoma could induce the same tumour in other chickens. At the time this discovery received little attention and eventually Rous abandoned his work on tumour viruses. Subsequent research over the next few decades clearly vindicated Rous' work, and in 1966 at the age of 85 he was awarded the Nobel Prize.

A great many viruses are now known to cause tumours. In man certain DNA viruses have been linked with particular cancers: hepatitis B virus with liver cancer, Epstein–Barr virus with Burkitt's lymphoma and nasopharyngeal carcinoma, and certain types of human papillomavirus (and probably herpesvirus) with cancer of the cervix. However, the recent developments in oncology have come mainly from the study of certain RNA viruses termed **retroviruses** because their RNA can be transcribed back into DNA.

All retrovirus particles (virions) have certain features in common. They consist of a regular external protein coat (capsid) and a central core of ribonucleoprotein containing linearly arranged single stranded RNA. In the RNA molecule there are typically three coding regions referred to as *gag*, *pol* and *env*, arranged in this order. The *gag* sequence codes for the protein component of the ribonucleoprotein core, the *pol* sequence codes for the enzyme reverse transcriptase and the *env* sequence codes for the protein coat. Certain retroviruses also possess an oncogene which, as we shall see confers on them the ability to produce neoplasms in infected animals. At either end of the RNA molecule are a series of repeated sequences. Events which occur when a retrovirus infects a cell are shown in Fig. 6.7.

Having lost its protein coat, the viral single-stranded RNA is copied by reverse transcriptase into double-stranded DNA. The latter possesses long terminal repeats (LTRs) at either end which are required for the integration of the viral DNA molecule into the host DNA. In its integrated form it is referred to as a **provirus** which replicates along with the host DNA. Transcription of the viral DNA produces viral genomic RNA and also mRNA for translation into proteins for new virus particles. In retroviruses possessing oncogenes, however, part of the mRNA is translated into certain proteins which are not components of the virus particle but are responsible for converting (transforming) an infected cell into a tumour

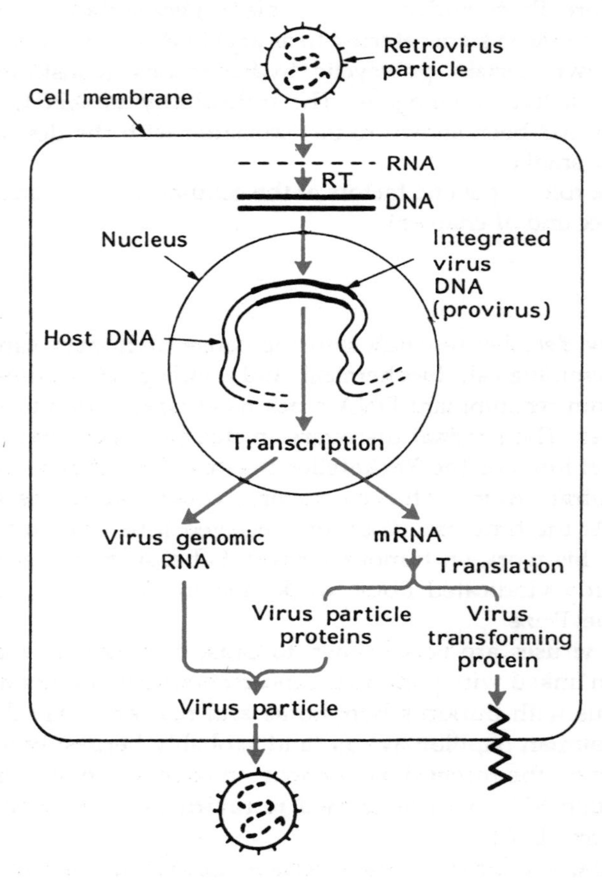

Figure 6.7 Sequence of events which occur when a retrovirus infects a cell (RT = reverse transcriptase).

cell. It now seems there are two types of tumour-causing (oncogenic) viruses. Those which are weakly oncogenic take months to produce tumours in infected animals and do not affect cells in culture. These viruses do not possess oncogenes. However, other viruses (like Rous sarcoma virus) do possess oncogenes and produce tumours in animals, often within a matter of days following infection. These viruses also transform cells in tissue culture, i.e. infected cells lose contact inhibition and become piled up in the culture dish instead of spreading out into regular flat layers of cells. The studies which have been so illuminating have come from studying cellular transformation by a technique referred to as transfection assay.

Transfection assay

This technique consists of extracting DNA from a solid tumour, or established cell lines from a tumour, and then applying it (in the presence of calcium phosphate which facilitates its uptake) to cultured mouse fibroblasts from a special cell line referred to as NIH/3T3. Under appropriate culture conditions a small proportion of the 3T3 cells will take up the tumour DNA and form a focus of transformed cells. DNA can then be extracted from the transformed cells and the procedure repeated on another culture, so producing secondary transformants. If DNA from a human bladder cancer, for example, is applied to the 3T3 cells, transformation occurs which implies that oncogenic DNA sequences have been taken up by these cells. Since secondary and even tertiary transformants can be produced, the transforming sequences must be relatively small in order to survive the serial extractions and manipulations involved which would have the effect of breaking up long strands of DNA. That these transformed cells do contain human DNA has been shown by extracting the DNA from the cells, cleaving it with a restriction endonuclease and then demonstrating hybridization with a radiolabelled probe (e.g. BLUR-8) containing *Alu* sequences. As we saw earlier, such a probe will identify DNA of human origin in a mixture with DNA of non-human origin because *Alu* sequences are specific for human DNA and occur so frequently in the genome that they are very likely to occur in the DNA sequence being sought (page 20). The transforming DNA can also be cloned and then characterized. The methodology is outlined in Fig. 6.8.

Oncogenes

By using viral oncogenic probes from animal retroviruses, human transforming DNA sequences have been found to be homologous to various viral oncogenes (v-onc). These have therefore been referred to as cellular oncogenes (c-onc) or proto-oncogenes. The homology between viral and corresponding cellular oncogenes has been confirmed by restriction enzyme analysis, heteroduplex analysis and sequencing.

It is believed that retroviral oncogenes originated from cellular genes which at some time in the past were picked up by retroviruses during their sojourn as proviruses. Relatives of human oncogenes occur widespread in the animal kingdom (even in yeast!) and would not have been retained over millions of years of evolution unless they had a very important function, but this is as yet not clearly

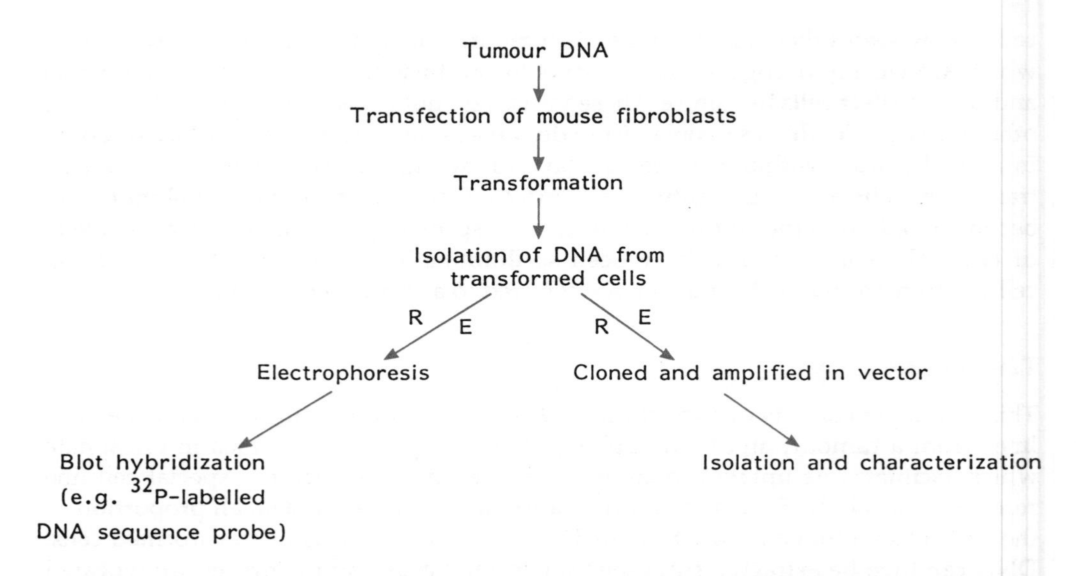

Figure 6.8 Methodology used to identify, isolate and characterize human transforming (oncogenic) DNA sequences (RE = restriction endonuclease).

understood. Some examples of cellular oncogenes which have been identified in humans with their chromosomal locations are given in Table 6.5.

So far activated oncogenes have been detected in a wide variety of malignant tumours including cancers of the bladder, breast, lung, colon and pancreas, neuroblastomas, Burkitt's lymphoma, certain leukaemias and sarcomas. In malignant tumours the activated oncogene is usually c-H-*ras* in bladder cancer, c-K-*ras* in many other cancers, c-N-*ras* in neuroblastomas and sarcomas, c-*myc* in Burkitt's lymphoma and certain cancers, and c-*abl* in chronic myeloid leukaemia. The same oncogene can thus be activated in several different tissues giving rise in each to a different type of tumour, and the same tumour may be associated with different activated oncogenes. Since cellular oncogenes occur normally the question arises as to how they become activated to produce overt cancer. The complete answer is not known though there are a number of possibilities which are not mutually exclusive. The evolution of a cancer cell is a multistep process and different mechanisms may operate at different stages in its evolution as well as in different tumours. There is even evidence that in many tumours more than one oncogene may be activated at the same time.

One possibility is a point mutation in a cellular oncogene which in some way then interferes with its expression. This was first reported independently by Reddy *et al.* (1982) and Tabin *et al.* (1982) in the oncogene c-H-*ras* in human bladder cancer. These investigators found that in the protein coded for by this oncogene, a protein kinase localized to the cell membrane and designated as p21 because it has a molecular weight of 21 000 daltons, at position 12 there was a substitution of valine for glycine. This mutation does not affect the level of expression of the oncogene but affects the structure of the oncogene-encoded protein. In fact subsequent work has shown that substitution of glycine at position 12 by any other amino acid has

Table 6.5 Some examples of viral oncogenes (v-onc) and the chromosomal location of homologous human cellular oncogenes (c-onc)

Origin	v-onc	c-onc location
Cat sarcoma		
Snyder–Theilin strain	*fes*	15
McDonough strain	*fms*	5
Chicken sarcoma		
Rous strain	*src*	20
Chicken erythroblastosis	*erb*-A	17
	erb-B	7
Myeloblastosis	*myb*	6
Myelocytomatosis	*myc*	8
Mouse sarcoma		
Moloney strain	*mos*	8
Mouse leukaemia		
Abelson strain	*abl*	9
Rodent sarcoma		
Harvey strain	H-*ras*-1	11
	H-*ras*-2	X
Kirsten strain	K-*ras*-1	6
	K-*ras*-2	12
	N-*ras*	1
Simian sarcoma	*sis*	22

the same transforming consequences. In some cancers mutations at position 61 in c-*ras* genes have also been reported. There must therefore be something special at these two positions in the p21 protein, perhaps affecting its secondary structure, which then in some way is responsible for malignant transformation. The *ras* family of genes are part of a complex signalling pathway involving hydrolysis and binding of guanosine triphosphate.

It is also possible that a cellular oncogene may become activated by a viral promoter when the relevant virus (by chance) becomes integrated into the DNA adjacent to the oncogene. This occurs in mice infected with the mammary tumour virus which activates a particular cellular oncogene and leads to the development of mammary cancer. A similar mechanism may account for T-cell leukaemia in humans due to a particular virus (HTLV) of which there are several types.

A third mechanism is by chromosomal rearrangement whereby a translocation results in a cellular oncogene being translocated to a site which is presumably adjacent to some form of promoter or enhancer (or alternatively the translocation releases the oncogene from being suppressed). The best examples (though there are several others) occur in Burkitt's lymphoma and chronic myeloid leukaemia. Burkitt's lymphoma is a cancer of the lymphoid tissues which is relatively common in African children but does occur in other parts of the world. It often presents with typical tumours of the jaw but later other sites and organs become affected. It is named after Denis Burkitt who first described the condition in the late 1950s. Burkitt's lymphoma responds well to chemotherapy and the prognosis is good if treatment is started early. Cell lines from patients with this disease have a translocation with part of chromosome 8, which carries the c-*myc* gene, being translocated usually to chromosome 14 close to the site of the immunoglobulin heavy-chain gene. A second disease is chronic myeloid leukaemia in which the c-*abl* gene on chromosome 9 is translocated to chromosome 22 (Fig. 6.9). This is a reciprocal translocation because part of the long arm of chromosome 22 is at the same time translocated to chromosome 9 which produces an apparently deleted chromosome 22 (22q−) which is referred to as the Philadelphia chromosome (Ph^1). A third example is the fusion of the α retinoic and receptor on chromosome 17 to a gene called PML on chromosome 15 which occurs in promyelocyte leukaemia. On this basis it has been suggested that derivatives of retinoic acid could be used clinically.

A fourth way in which a cellular oncogene may be activated is by chemical carcinogens, probably through the induction of a point mutation in the relevant oncogene. A fascinating example of this comes from studies of mutations of the p53 gene in hepatocellular cancer. This gene can act as a dominant oncogene and also has characteristics of a tumour suppressor gene. It is mutated in about half of all types of cancer arising from a wide spectrum of tissues and a very wide range of mutations has been found. Mutations of the gene were studied in Mozambique and Qidong, China, where chronic infection with hepatitis B virus plus dietary exposure to the carcinogenic mould aflatoxin B_1 combine to give a high incidence of

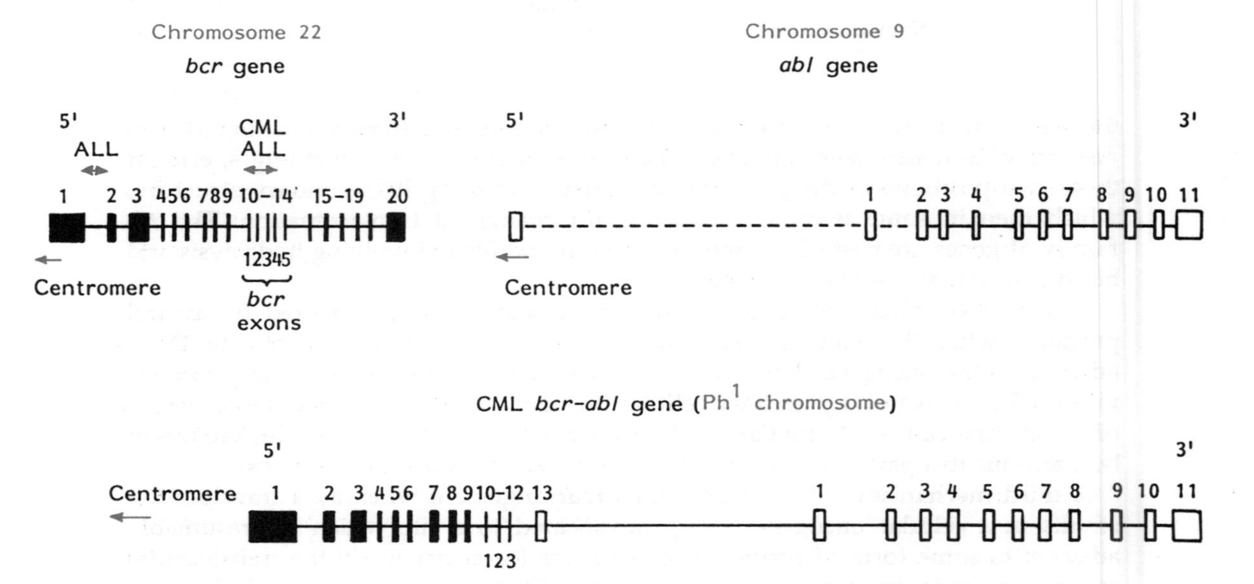

Figure 6.9 Normal c-*abl* and *bcr* genes on chromosomes 9 and 22, respectively, and the chimeric *bcr–abl* gene associated with chronic myeloid leukaemia (CML). The boxes represent exons (DNA sequences that are included in the mRNA); the number above each box denotes the exon number. The relative sizes of the exons and introns and the number of exons may not be exactly as shown. Transcription starts at the 5' end of the genes. The dashed line in the c-*abl* gene represents the DNA segment in which breakpoints occur. The symbol, ↔, above the *bcr* gene, denotes the sections of this gene in which breaks occur in CML or acute leukaemia, as designated. (From Kurzrock *et al.*, 1988, with permission.)

hepatocellular carcinoma. Ten of the first 13 mutations described were guanine-to-thymine transversions that all occurred at the third base of codon 249 resulting in a serine to arginine change. This contrasts with the spread of mutations usually found. Studies of hepatocellular carcinoma around the world showed they could be divided into two sets. In high incidence areas (>40 000 cases per 100 000) the G–T mutation was common but in other areas with a lower incidence (<40 000 cases per 100 000) a wider spectrum of mutations is found. This strongly suggests that the G to T change can be attributed to the dietary aflatoxin, a finding which is supported by chemical evidence that this toxin has powerful mutagenic activity.

One final point arises. Because of the discovered homologies between viral oncogenes and cellular oncogenes, fears have been expressed that cancer may be infectious. However, except for certain DNA viruses (such as hepatitis B which may cause liver cancer) and the very rare T-cell leukaemia-lymphoma virus, cancers are not infectious in the usual sense of the word. It seems that the pathogenesis of most cancers lies not in infection but in the activation of cellular oncogenes which are already carried by each one of us.

Some time has been spent on the possible mechanism of oncogene activation because of the fundamental importance of this process in the pathogenesis of cancer. The identification of oncogene encoded proteins is also of fundamental importance. Now that many of these have been identified we have some information about the classes of genes involved. They have been summarized by Cline (1994). Firstly, there are molecules involved in transferring signals for cell growth from the cell membrane to the nucleus (examples are *ras* and c-*abl*). Secondly there are molecules that activate transcription by binding to specific sequences of DNA (*myc* is an example). The third group comprises genes which also alter transcription and lead to tissue differentiation and are recognized because they contain particular sequences known as homeobox domains. The fourth group consists of genes involved in apoptosis (*bcl*). The fifth group contains tumour suppressor genes but these will be dealt with in the next section. Table 6.6 summarizes genetic abnormalities in leukaemia.

Tumour suppressors

The discovery of tumour suppressors and their relevance to inherited common cancers is well illustrated by a study of retinoblastoma. Retinoblastoma is a malignant tumour of the eye which presents in early childhood. Most unilateral cases are sporadic and believed not to be genetic, but bilateral cases are often inherited. The tumour arises by two mutational steps at the retinoblastoma (rb) locus which is located on the long arm of chromosome 13 (13q14). Hereditary cases have a germ-line mutation which is present in all the somatic and germ-line cells and which can be transmitted from one generation to another. Tumour formation only results when a second, and this time somatic, mutation occurs in a retinal cell at the rb locus on the other homologous chromosome. In sporadic cases both mutations occur as chance events in the same retinal cell. This so-called 'two-hit' hypothesis to account for certain familial cancers was first proposed independently by de Mars and Knudson several years ago, but before the advent of recombinant DNA technology there was no way of really testing it.

Table 6.6 Genetic abnormalities in leukaemia

Type of gene	Function	Molecular alteration	Chromosomal abnormality
ras	Signal transduction	N-*ras* mutations	None
Tyrosine kinases	Membrane signal transductions	Fusion c-*abl* to *bcr*	t(9;22)(q34;q11)
Transcriptional control element	Gene transcription	Fusion *myc* to immunoglobulin genes	t(8;14)(q24;q32)
		Fusion E2A to PBX or HLF	t(1;19)(q23;p13 t(17;19)(q22;p13)
		Fusion *tal*-1 to T cell receptor genes	t(1;14)(p32;q11)
		Fusion *ly*-1 to T cell receptor genes	t(7;19)(q35;p13)
		Fusion *Ttg*-1 or *Ttg*-2 to T cell receptor genes	t(11;14)(p15;q11) t(11;14)(p13;q11)
Homeodomain	Differentiation and gene transcription	Fusion of HOX-11 to T cell receptor genes	t(10;14)(q24;q11)
		HRX translocation	t(11q23)
Receptor	Differentiation	Fusion of α-retinoic acid receptor gene to PML	t(15;17)(q21;q21)
bcl	Apoptosis	Fusion *bcl*-2 to immunoglobulin genes	t(14;18)(q32;q21)
	Cell cycle control	Fusion *bcl*-1 to immunoglobulin genes	t(11;14)(q13;q32)
	Inhibition of gene transcription	Fusion *bcl*-3 to immunoglobulin genes	t(14;19)(q32;q13)

Adapted from Cline (1994).

There are a number of possible ways by which a recessive mutation of the rb locus on one chromosome in a retinal cell might be revealed and thus lead to development of a tumour. Some of these are illustrated in Fig. 6.10. They include mitotic non-disjunction whereby the chromosome bearing the normal gene is lost during an error of cell division in a retinal cell (Fig. 6.10b), mitotic non-disjunction with subsequent reduplication of the mutant chromosome (Fig. 6.10c), mitotic recombination between the rb locus and the centromere resulting in homozygosity at the rb locus as well as the rest of the distal part of the chromosome (Fig. 6.10d), small deletions of the normal (+) allele at the rb locus (Fig. 6.10e and f) and a point mutation of the normal gene (Fig. 6.10g). Cavenee and colleagues (1983), along with others, applied recombinant DNA technology to this problem including karyotyping tissue from retinoblastoma cells. Using cloned hybridization probes, RFLPs at loci flanking the rb locus were studied in constitutional tissues (leucocytes and fibroblasts) as well as in tumour cells removed at the time of operation. In some cases which proved to be heterozygous in constitutional tissues for the DNA markers, it was found in tumour tissue that at the polymorphic loci only **one** of the codominant alleles was present. In some cases this was associated with loss of a

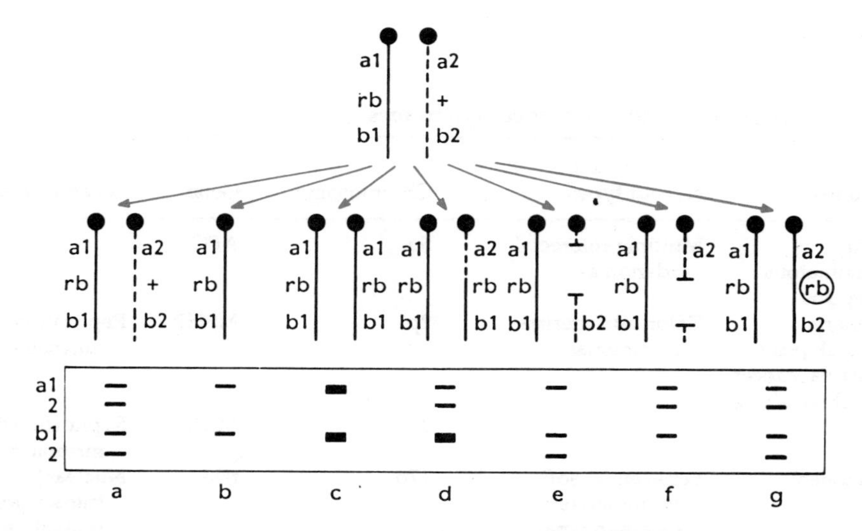

Figure 6.10 Above: possible ways in which a recessive mutation of the retinoblastoma (rb) locus might be revealed. Below: The expected associated DNA fragments on gel electrophoresis using hybridization probes for RFLPs at a locus proximal (alleles a1 and a2) and at a locus distal (alleles b1 and b2) to the rb locus. (a) Represents an individual who is heterozygous in constitutional tissues at the two polymorphic loci flanking the rb locus. (b) to (g) Represent the possible findings in retinal tissue (for details see text).

chromosome 13 presumably carrying the normal allele (Fig. 6.10b) or with apparent duplication of the mutant chromosome (Fig. 6.10c). In other cases in tumour cells heterozygosity was retained for markers proximal to the locus but homozygosity for markers distal to the locus (Fig. 6.10d). Finally, in yet other cases, there was evidence from DNA markers of sub-microscopic deletions in tumour tissues (Fig. 6.10e and f).

Since the isolation of the rb gene its product has been found to be a nuclear phosphoprotein with DNA binding activity which forms an inactivating complex with several tumour-producing proteins. In fact it is vitally important in the control of a very wide range of tumours and the reasons why the inherited form should be confined to retinal cells is quite unknown.

The general experimental principle outlined above, now entitled **loss of heterozygosity** or LOH, has been widely used in cancer studies and provides a method for mapping chromosomal areas potentially containing tumour suppressor genes in different tumour types. This is particularly useful in clinically and molecularly complex cases such as ovarian cancers where frequent loss of alleles from chromosome 6 has been observed. Experimentally, this particular task has become much easier with the arrival of PCR. Many alleles differing in length can rapidly be analysed even using poor quality DNA from tumours. However, the interpretation of results is often complicated by the impossibility of obtaining uncontaminated tumour cells.

Table 6.7 Dominantly inherited cancer syndromes

Syndrome	Predominant tumour types	Chromosome	Gene	Location/function
Familial adenomatous polyposis	Multiple colorectal adenomas	5q	APC	
Hereditary non-polyposis colorectal cancer (Lynch syndrome)	Colon and uterine carcinomas	2p	MSH2	Replication fidelity/ mismatch repair
		3	MLH1	Replication fidelity/ mismatch repair
Li–Fraumeni	Leukaemia, soft tissue sarcoma, osteosarcoma, brain, breast and adrenal cortical tumours	17p	p53	Nuclear/ transcriptional regulation
Multiple endocrine neoplasia type 2	Medullary thyroid carcinoma, phaeochromocytoma	10	Ret	Receptor tyrosine kinase
Neurofibromatosis type 1	Multiple peripheral neurofibromas	17q	NF1	Cytoplasmic GTPase activating protein
Neurofibromatosis type 2	Central schwannomas and meningiomas	22q	NF2	Cytoskeletal-membrane link
Retinoblastoma	Retinoblastomas, osteosarcomas	13q	rb	Nuclear/transcriptional regulation
Von Hippel–Lindau	Renal cell carcinoma	3p		
Wilms tumour	Nephroblastoma	11p	WT1	Nuclear/transcriptional regulation
Tuberous sclerosis type 2		16p	TSC2	G protein signalling
Early onset familial breast cancer	Breast and ovary	17q	BRCAI	Zincfinger/DNA binding
		13q	BRCAII	
Familial melanoma	Melanoma	9p		
Gorlin syndrome	Basal cell carcinoma of the skin	9q		
Multiple endocrine neoplasia type 1	Parathyroid, endocrine, pancreas and pituitary tumours	11q		

Great progress has been made in the isolation of dominantly inherited genes predisposing to tumours and many of these require a second hit for a tumour to be formed. Others have been mapped but not yet isolated, although progress is likely to be rapid. The current situation is shown in Table 6.7. Most recently a gene for early onset breast and ovarian cancer has been isolated, and a second locus mapped to chromosome 13.

CANCER FAMILY SYNDROMES

Three of the examples in Table 6.7 are particularly interesting in that they help solve the mystery of families in which there is a strong clustering of cancers, but of different types. In the first example, Li–Fraumeni syndrome, the inherited mutation is in the p53 gene which has been implicated in many tumour types. The second example is Lynch cancer family syndrome II in which members of the families are predisposed to many forms of cancer but particularly carcinomas of the colon or endometrium (hence the alternative name of hereditary non-polyposis colon cancer, or HNPCC, which we shall use here). During the original linkage studies used to map this gene (there subsequently turned out to be two loci, one on chromosome 2 and one on chromosome 3) it was noted that the CA repeat markers being analysed often showed alterations in the number of repeats between generations. Such fresh mutations are occasionally seen in all linkage studies but at nothing like the rate observed in these families. This instability was observed in markers from all over the genome suggesting that there may be an underlying mutation in the molecular apparatus which normally corrects these errors. The mismatch repair pathway is best defined in *E. coli* where four gene products have been identified, MutH, MutS, MutL and MutU (uvrD). MutS protein binds to mismatched nucleotides in DNA recombination intermediates and MutL interacts with the bound MutS and MutH which carries out the excision needed for repair. The human homologues of MutS and MutL have been isolated and shown to map to chromosome 2 and 3 respectively. Missense mutations have been found in both genes in HNPCC families proving these are the underlying genes in this disorder. It is believed that mutations of these genes could be carried by as many as one in 200 people and possible screening strategies based on the genomic instability could be devised. Presymptomatic screening would be particularly appropriate because it would identify those members of the population who should have regular screening for early signs of colon cancer. With early detection the cancer can be cured. However, in many of the families involved genetic diagnosis is, frustratingly, not immediately possible because with only a few family members available it is not possible to distinguish whether the gene is of the chromosome 2 or chromosome 3 type or at least one other type which has not yet been mapped.

Even from this very much simplified discussion of a very complex field it will be appreciated that an understanding of the molecular genetic basis of cancer is likely to be of clinical value in three main areas. Firstly, it provides refined methods for classifying cancers. Secondly, it is possible to develop relatively simple non-invasive methods for diagnosing cancer and predicting cancer risks. Thirdly, it may lead to the development of rational therapies based on controlling oncogene activity and antagonizing the effects of their protein products.

SUMMARY

- Genes have been identified which cause common genetic disorders in rare sub-sets of families.
- Association studies, using DNA polymorphisms, have identified some genes likely to make a genetic contribution in some common disorders. Examples include apolipoprotein AI in atherosclerosis and coronary heart disease and apolipoprotein E4 in Alzheimer disease.
- The products of several oncogenes have been identified, and they include growth factors, hormone receptors and DNA binding factors.
- Mechanisms of oncogene activation include point mutation (spontaneous or chemically induced), a viral promoter and chromosomal translocation.
- In inherited cancers, there is usually a tumour suppressor, or anti-oncogene involved, whose normal function is to hold uncontrolled cell growth in check. Several of these have been identified. There are multiple changes during the formation of a tumour and the development of cancer is a multi-step process.

REFERENCES AND FURTHER READING

Textbooks and review articles

Bishop JM. Molecular themes in oncogenesis. *Cell* 1991; **64**: 235–238
Cline M. The molecular basis of leukaemia. *N Engl J Med* 1994; **330**: 328–336
Cooper GM. Cellular transforming genes. *Science* 1982; **218**: 801–806
Doll R, Peto R. *The Causes of Cancer*. Oxford: Oxford University Press, 1981
Humphries SE. Life style, genetic factors and the risk of heart attack: the apolipoprotein B gene as an example. *Biochem Soc Trans* 1993; **21**: 569–582
Kosik KS. Alzheimer's disease: a cell biological perspective. *Science* 1992; **256**: 780–783
Krontiris TG. The emerging genetics of human cancer. *N Engl J Med* 1983; **309**: 404–409
Kurzrock R, Gutterman JU, Talpaz M. The molecular genetics of Philadelphia chromosome positive leukemias. *N Engl J Med* 1988; **319**: 990–998
Prusiner SB. Molecular biology of prion diseases. *Science* 1991; **252**: 1515–1521
Rabbitts TH. Translocations, master genes and differences between origins of acute and chronic leukemias. *Cell* 1991; **76**: 641–644
Solomon E, Borrow J, Goddard AD. Chromosome aberrations and cancer. *Science* 1991; **254**: 1153–1160
Trent JM, Metzler PS. The last shall be first. *Nature Genetics* 1993; **3**: 101–102
Wallace DC. Mitochondrial diseases: genotype versus phenotype. *TIG* 1993; **9**: 128–133
Weinberg RA. Tumor suppressor genes. *Science* **254**: 1138–1146
Weiss R, Teich N, Varmus H, Coffin J (Eds). *RNA Tumor Viruses*. Cold Spring Harbor Lab., 1982

Research publications

Arnold A, Cossman J, Bakhshi A, Jaffe ES, Waldmann TA, Korsmeyer SJ. Immunoglobulin-gene rearrangements as unique clonal markers in human lymphoid neoplasms. *N Engl J Med* 1983; **309**: 1593–1599

Bell GI, Karam JH. The polymorphic locus flanking the human insulin gene: is there an association with diabetes mellitus? In: Caskey CT, White RL, eds. *Recombinant DNA: Applications to Human Disease*, Cold Spring Harbor Lab. (Banbury Report 14), 1983; pp. 317–324

Bell GI, Horita S, Karam JH. A polymorphic locus near the human insulin gene is associated with insulin-dependent diabetes mellitus. *Diabetes* 1984; **33**: 176–183

Cavenee WK, Dryja TP, Phillips RA, Benedict WF, Godbout R, *et al*. Expression of recessive alleles by chromosomal mechanisms in retinoblastoma. *Nature* 1983; **305**: 779–784

Corder EH, Saunders AM, Strittmatter WJ, Schmechel DE, Gaskell PC, Small GW, Roses AD, Haines JL, Pericak-Vance MA. Gene dosage of Apolipoprotein E type 4 allele and the risk of Alzheimer's disease in late onset families. *Science* 1993; **261**: 921–923

Emery AEH, Anand R, Danford N, Duncan W, Paton L. Aryl hydrocarbon hydroxylase inducibility in patients with cancer. *Lancet* 1978; **i**: 470–472

Hamer DH, Hu S, Magnuson VL, Hu N, Pattatucci AML. A linkage between DNA markers on the X chromosome and male sexual orientation. Deletions of muscle mitochondrial DNA in patients with mitochondrial myopathies. *Science* 1993; **261**: 321–327

Harris CC, Hollstein M. Clinical implications of the p53 tumour suppressor gene. *N Engl J Med* 1993; **329**: 1318–1326

Holt IJ, Harding AE, Morgan-Huges JA. *Nature* 1988; **331**: 717–719

Kadowaki T, Kadowaki H, Mori Y, Tobe K, Sakuta R. A subtype of diabetes mellitus associated with a mutation of mitochondrial DNA. *N Engl J Med* 1994; **330**: 962–968

Lucassen AM, Julier C, Beressi J-P, Boitard C, Frognel P, Lathrop M, Bell JI. Susceptibility to insulin dependent diabetes mellitus maps to a 4.1 kb segment of DNA spanning the insulin gene and associated VNTR. *Nature Genetics* 1993; **4**: 305–310

Mandrup-Poulsen T, Owerbach D, Mortensen SA, Johansen K, Meinertz H, *et al*. DNA sequences flanking the insulin gene on chromosome 11 confer risk of atherosclerosis. *Lancet* 1984; **i**: 250–252

Matheson JAB, Matheson H, Anderson SA. Familial ovarian cancer. How rare is it? *Roy Coll Gen Pract* 1981; **31**: 743–745

Morel PA, Dorman JS, Todd JA, McDevitt HO, Trucco M. Aspartic acid at position 57 of the HLA DQβ chain protects against type I diabetes: a family study. *Proc Natl Acad Sci USA* 1988; **85**: 8111–8115

Reardon W, Ross RJM, Sweeney M, Luxon LM, Pembrey ME, Harding AE, Trembath RC. Diabetes mellitus associated with a pathogenic point mutation in mitochondrial DNA. *Lancet* 1992; **340**: 1376–1379

Reddy EP, Reynolds RK, Santos E, Barbacid M. A point mutation is responsible for the acquisition of transforming properties by the T24 human bladder carcinoma oncogene. *Nature* 1982; **300**: 149–152

Rosen DR, Siddiqu T, Patterson D, *et al*. Mutations in Cu/Zn superoxide dismutase gene are associated with familial amyotrophic lateral sclerosis. *Nature* 1993; **362**: 59–62

Rotter JI, Rimoin DL. Diabetes mellitus. In: Emery AEH, Rimoin DL, eds. *Principles and Practice of Medical Genetics*, 2nd edn. Edinburgh and London: Churchill Livingstone, 1990, pp. 1521–1558

Stritmatter WJ, Saunders AM, Schmechel D, Pericak-Vance MA, Enghild J, Salvesen GS, Roses AD. Apolipoprotein E: high affinity binding to β-amyloid and increased frequency of type 4 allele in late-onset familial Alzheimer disease. *Proc Natl Acad Sci USA* 1993; **90**: 1977–1981

Tabin CJ, Bradley SM, Bargmann CI, Weinberg RA, Papageorge AG, *et al*. Mechanism of activation of a human oncogene. *Nature* 1982; **300**: 143–149

Todd JA, Bell JI, McDevitt HO. HLA-DQβ gene contributes to susceptibility and resistance to insulin dependent diabetes mellitus. *Nature* 1987; **329**: 599–605

Valdheim CM, Rimoin DL, Rotter JI, Emery AEH, Rimoin DL (eds). Diabetes mellitus, in *Principles and Practice of Medical Genetics*, 2nd edn. Edinburgh and London: Churchill Livingstone, 1990, pp. 1521–1558

Wallace DC. Mitochondrial DNA deletion associated with Leber's hereditary optic neuropathy. *Science* 1988; **242**: 1427–1430

Chapter 7
Prevention of genetic disease

Most genetic diseases are serious and as yet very few are treatable. The only approach is therefore prevention which in **genetic** terms can be considered at three levels. Firstly, there is **primary** prevention by which is meant prevention at the level of the gene or chromosome by reducing the frequency of new mutations. Theoretically this might be achieved by the avoidance of environmental mutagens or by manipulating the genome in some way so that it is resistant to such mutagens. However, apart from the avoidance of radiation, primary prevention is not remotely feasible at the present moment and in any case we now understand that a lot of instability is an inevitable result of the organization of the genome, particularly the presence of repeated sequences. By **secondary** prevention is meant the avoidance of aetiologically significant factors in the environment in those disorders which have a multifactorial basis and where environmental factors play a large part in causation. Vitamin supplementation and general dietary improvement have been claimed to reduce significantly the recurrence of neural tube defects. However, in almost all other common congenital malformations no aetiologically important environmental factors have so far been identified which would revolutionize the approach to prevention in these disorders. Finally, there is **tertiary** prevention by which is meant the prevention of genetic disease achieved by ascertaining those individuals in the population who are at risk of transmitting a serious genetic disorder to their offspring, and then offering them genetic counselling and prenatal diagnosis with selective abortion of affected fetuses when this is possible.

GENETIC COUNSELLING

Genetic counselling is essentially a process of communicating information which falls roughly into two main areas: firstly, information about the disease itself – its severity and prognosis, whether or not there is any effective therapy, the genetic mechanism which caused the disease and the risks of recurrence; secondly, information on the available options open to a couple who find the risks unacceptably high and this may include discussions of contraceptive methods, adoption, prenatal diagnosis and abortion and artificial insemination by donor (AID).

Genetic risks form an important part of counselling. They are highest in unifactorial disorders. In some circumstances, e.g. in autosomal dominant disorders of late onset, it may be necessary to determine if an otherwise healthy individual carries the mutant gene and is therefore a preclinical case. In X-linked recessive disorders a female relative may have to be be tested to see if she is a carrier. In cytogenetic disorders the risks, depend on the particular disorder. In Down's syndrome, due to trisomy-21, the risks depend on the mother's age, being about one in 2000 at the age of 20, one in 800 at the age of 30, one in 100 at the age of 40, and thereafter increase considerably. Details of genetic risks are given in several excellent texts. They are an important factor in influencing parents' decision making, and along with other factual information form the basis of genetic counselling. However, nowadays more consideration is increasingly being given to the psychological aspects of counselling.

Here the emphasis is on fully appreciating the psychological impact of genetic disease on the individual couple, recognizing the various stages of the so-called 'coping process' and tailoring counselling accordingly. Thus there has been movement away from **purely** fact-oriented to more person-oriented counselling. This change of emphasis has been brought about by the recognition that genetic disease *per se* frequently has profound psychological effects on parents, often with long-term consequences. Perhaps the most important factor has been the realization that couples given genetic counselling may opt for a course of action which may be at variance with what the counsellor might have regarded as 'reasonable' and 'responsible'. The point is that the final decision must always rest with the couple themselves, and their course of action will be influenced by a great many personal, social and psychological factors, quite apart from genetic considerations. However, the great improvement in the accuracy of risk estimation possible with DNA technology makes it easier for parents to make well-informed decisions. Since the counsellor's role must always be to help couples reach decisions which are the best for themselves, genetic counselling should never be directive.

PRENATAL DIAGNOSIS

To the couple who decide they could not accept an affected child, prenatal diagnosis has been a major development and has removed much of the uncertainty from genetic counselling. Currently there are four techniques which can be used to diagnose an affected fetus sufficiently early in pregnancy for termination to be possible should the fetus prove to be affected: ultrasonography, fetoscopy, amniocentesis and chorion biopsy (Table 7.1). Although far from routine, preimplantation embryo diagnosis is now a practical possibility.

Ultrasonography, or **sonography** for short, involves the use of high-frequency sound waves which are converted to visual signals on a television screen. It is a non-invasive method which causes no risks to the mother or fetus but requires somewhat expensive equipment and considerable expertise in the accurate interpretation of the results. It is used routinely for the obstetric indications such as placental localization, fetal viability, fetal skull and limb measurements to determine growth rate and maturity, and for the diagnosis of multiple pregnancies. It has also found use in the prenatal diagnosis of various congenital abnormalities where the defects cannot be easily visualized directly and which are not associated

Table 7.1 Techniques and some indications for prenatal diagnosis

Technique	Indication	Examples
First trimester		
(a) Chorion biopsy	Fetal sexing	
	Chromosome abnormalities	
	Metabolic disorders	
	DNA studies	
Second trimester		
(b) Ultrasonography	CNS abnormalities	Spina bifida, anencephaly, microcephaly, hydrocephaly
	Renal abnormalities	Infantile polycystic kidneys
		Renal agenesis
	Gastrointestinal abnormalities	Intestinal atresia, omphalocele
	Short-limbed dwarfism	Achondrogenesis
	Limb defects	Phocomelias
	Cardiac abnormalities	Septal defects
(c) Fetoscopy	Morphological abnormalities	Facial clefts, peripheral limb defects
	Blood sampling	Haemophilia, haemoglobinopathies
	Biopsy (skin)	Epidermolysis bullosa, congenital ichthyosis
(d) Amniocentesis	Amniotic fluid alphafetoprotein	Anencephaly, spina bifida
	Amniotic fluid cells	Fetal sex, chromosome abnormalities, metabolic disorders, DNA studies

with any chromosomal or biochemical defect which could be detected in amniotic fluid or its contained cells. Abnormalities which have been detected in this way include CNS abnormalities (microcephaly, hydrocephaly), renal abnormalities (infantile polycystic kidneys, renal agenesis), gastrointestinal abnormalities (intestinal atresia, omphalocele), various forms of short-limbed dwarfism (e.g. achondrogenesis), peripheral limb defects (e.g. phocomelias) and even certain cardiac abnormalities such as ventricular septal defects. Neural tube defects have also been detected by sonography, but alphafetoprotein levels in amniotic fluid (page 132) have generally found more favour because of the greater sensitivity.

The fetus may also be visualized directly by inserting a very fine fibre optic telescope (**fetoscope**) through the anterior abdominal wall into the uterine cavity. In this way it is possible to diagnose, for example, facial clefts and peripheral limb defects (e.g. Ellis–van Creveld syndrome, ectrodactyly). Using a fetoscope it is also possible to insert a needle into the cord and so obtain samples of fetal blood, a technique which, until recently, was largely used for the prenatal diagnosis of haemophilia and haemoglobin variants, though this approach has mainly been overtaken by recombinant DNA technology. It has even been possible to use the technique to take a skin biopsy from the fetus in order to detect certain inherited dermatological conditions (such as epidermolysis bullosa and congenital ichthyosis).

The technique most widely used for prenatal diagnosis is that of trans-abdominal **amniocentesis,** whereby a small volume (10–20 ml) of amniotic fluid is aspirated through a needle inserted into the amniotic cavity, care being taken to avoid the placenta, the position of which has previously been defined by ultrasonography (Fig. 7.1). Amniocentesis cannot be carried out before the 12th to 14th week of gestation because there is relatively little fluid and the uterus is below the pelvic brim which makes the procedure more difficult and hazardous. In practice it is usually carried out around 16 to 18 weeks of gestation. In experienced hands there is virtually no risk to the mother though there is a possibility of the procedure producing an abortion, variously estimated to be from 1 to 2% above the normal incidence of spontaneous abortions at this time of pregnancy.

The cells in the amniotic fluid are of fetal origin, being derived from fetal skin and amnion. Most are dead squames but a proportion are viable and will grow in tissue culture. In this way the chromosome constitution, including the sex, of the

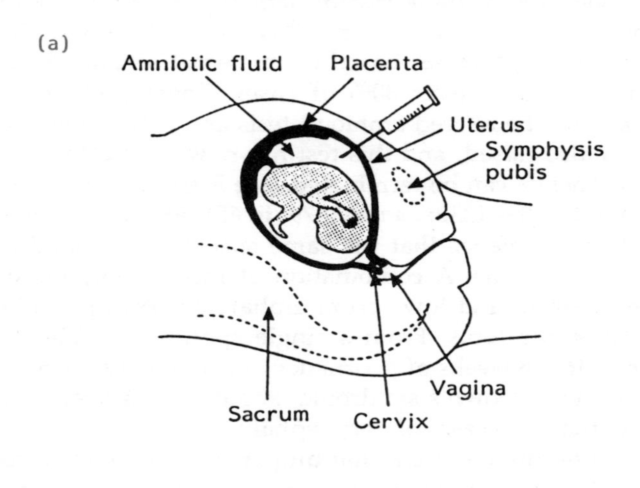

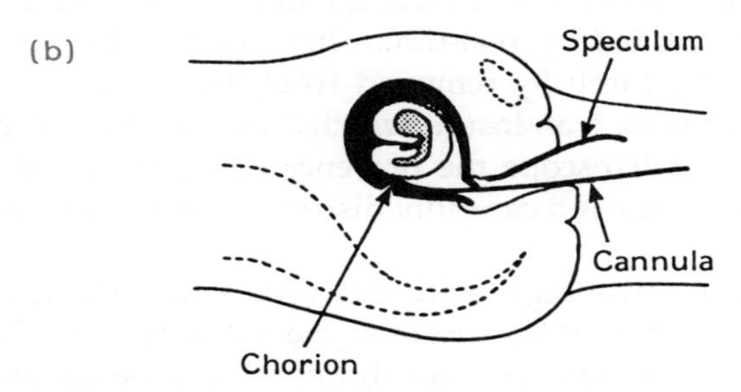

Figure 7.1 Technique of: (a) transabdominal amniocentesis, (b) chorion biopsy.

fetus can be determined. In X-linked disorders in which the affected male fetus cannot be diagnosed prenatally, if the mother is at high risk of having an affected son it is possible to sex the fetus and offer the mother selective abortion of any male fetus. In this way she can be guaranteed a daughter who will not be affected. By studying the fetal karyotype it is also possible to diagnose various chromosomal abnormalities, the most frequent indication being the risk of Down's syndrome in pregnancies of older mothers. Cultured amniotic fluid cells may also be used to diagnose various inborn errors of metabolism provided the responsible defect is expressed in such cells. So far some sixty biochemical disorders have been diagnosed in this way. Disorders in which the specific enzyme deficiency cannot be demonstrated in epithelial tissues but only in other tissues (e.g. phenylalanine hydroxylase is only present in liver cells) cannot therefore be detected *in utero* by this technique (e.g. phenylketonuria). It usually takes up to two weeks in order to carry out chromosome studies on cultured amniotic fluid cells but it may take as long as six to eight weeks for there to be sufficient cells for enzyme studies.

One of the most important indications for prenatal diagnosis, at least in the United Kingdom, is the detection of neural tube defects which is possible from determining the level of alpha-fetoprotein in amniotic fluid around the 18th week of gestation, and the test is offered to any mother who has had a previously affected child. However, over 90% of these malformations are born to mothers who have never had a previously affected child. This has led to the development of population screening programmes based on maternal serum levels of alphafetoprotein which are raised in over 80% of cases. Those mothers found to have elevated levels can then be offered amniocentesis and if the amniotic fluid level of alphafetoprotein is also raised, and this test has now been refined in a number of ways, then the pregnancy can be terminated (see Reports of the UK Collaborative Studies: *Lancet* 1977; i: 1323–1332; *Lancet* 1979; ii: 651–662; *Lancet* 1981; ii: 321–324).

It is a fortunate convenience that the same specimen of blood can be used to screen for Down's syndrome. A combination of raised chorionic gonadotrophin, low unconjugated oestriol and low serum alphafetoprotein provide a marker for Down's syndrome in the fetus. From a single specimen of blood taken from a pregnant woman at 16–18 weeks of pregnancy, a combined screen can detect about 60% of pregnancies with Down's syndrome, about 90% of pregnancies with open spina bifida and virtually all cases of anencephaly.

More recently the technique of **chorion biopsy** has been introduced for prenatal diagnosis. Essentially this consists of inserting a flexible cannula through the cervix into the uterine cavity guided by real-time ultrasound or by direct visualization (Fig. 7.1). Material is then carefully removed from the chorion – usually near the edge of the so-called chorion frondosum which will later form the placenta. With the help of a dissecting microscope the presence of chorionic villi (which are of fetal origin) can be confirmed and carefully dissected away from any other material which may be present.

There are considerable advantages with this technique. Firstly, it can be carried out earlier in pregnancy than amniocentesis, the optimum time being around 10 weeks of gestation, which means that if an abortion has to be carried out this can be done much earlier and will therefore cause less psychological trauma. Secondly, under appropriate conditions, chromosome and enzyme studies can be performed on **uncultured** villus material which saves considerable time and expense. But there are also problems. If there is a risk of a neural tube defect, amniocentesis for alphafetoprotein determination may have to be performed later in any event, though maternal serum alphafetoprotein and sonography may suffice. Also, certain chromosome abnormalities (such as trisomy-16) may be detected at this stage of pregnancy which would normally abort spontaneously. Finally, the fetal loss associated with the procedure may prove to be quite high. In a British randomized trial, 4.6% fewer of the women randomized to CVS rather than aminiocentesis, had a surviving child. Nevertheless, for fetal sexing, the detection of Down's syndrome and inborn errors of metabolism and for DNA studies, this may well become the preferred method for obtaining fetal material.

THE NEW TECHNOLOGY AND THE PREVENTION OF GENETIC DISEASE

Recombinant DNA technology has introduced many innovations into the field of prevention. Most of these developments stem from the fact that the new technology provides a means of detecting defects at the molecular level in genes normally expressed only in relatively inaccessible tissues. For example, the dystrophin gene is only expressed in muscle tissue but is, of course, present in all other cells of the body including peripheral blood leucocytes, skin fibroblasts and therefore chorion villus biopsy (CVB) cells. The new technology provides methods for detecting a mutant dystrophin gene in any of these other tissues and without the material having to be cultured. Perhaps of greater importance is the fact that a genetic defect may be detectable even when the basic biochemical abnormality is unknown. In this case the technology is exploited in finding close linkage between the mutant gene in question and a DNA polymorphism.

Genetic linkage

As discussed earlier, gene loci are said to be linked when they are carried on the same chromosome pair, and their distance apart is measured by the frequency with which crossing over (recombination) occurs between them during gametogenesis (page 59). If in a particular individual two alleles are on the same chromosome they are said to be in the **coupling** phase, but if they are on different homologous chromosomes they are said to be in the **repulsion** phase. In families in which a genetic disorder and a marker trait (e.g. a DNA polymorphism) occur over three or more generations, then the linkage phase of individuals in intermediate generations may be obvious from inspection of the pedigree and the amount of crossing-over (usually referred to as the recombination fraction or θ) can be determined quite simply. An example is given in Fig. 7.2.

Let us assume that the disease in question is inherited as an autosomal dominant trait, the disease gene being represented as D and its normal (recessive) allele as d, and the DNA polymorphism consists of two codominant alleles A and B. We see that II-1 must be heterozygous at both loci and the D and B alleles are in coupling since her father was affected and only carried the B allele and her mother only

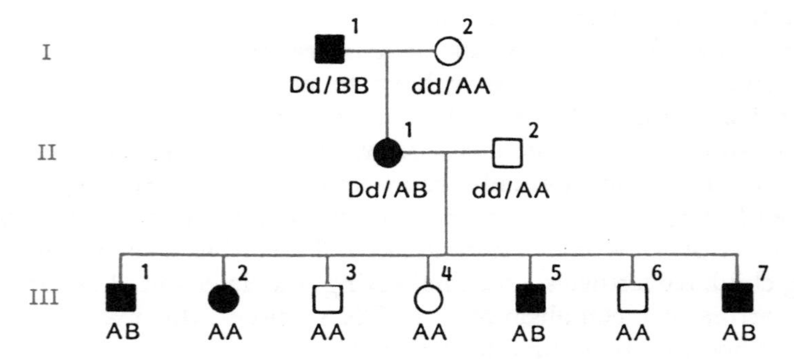

Figure 7.2 Family in which genes for a disease (D) and a DNA polymorphism (A, B) are segregating. In the third generation only the phenotypes for the DNA polymorphism are indicated.

carried the A allele. Therefore in the third generation individual III-2 must be a recombinant and therefore the recombination fraction is one out of seven or 0.14, so that the two loci would appear to be about 14 map units or centiMorgans (cM) apart. But rarely are human pedigrees so obliging. Often it is necessary to calculate the most likely value of the recombination fraction making various assumptions about the linkage phase in various family members, and the data from several families are then combined. For convenience this likelihood (or odds) is expressed as its logarithm, the $\log_{10}$ of the odds being referred to as the **lod score**. Lod scores for different values of θ are calculated for each family and then these are added together and the sum of the lod scores plotted against various values of θ. The maximum likelihood estimate of θ is the value of θ which corresponds to the peak of the curve. Thus if the peak of the curve was 2.805 for a value of θ of 0.05, then the probability of linkage at this distance corresponds to the antilog of 2.805 or 638:1. The 95% probability limits of the maximum likelihood estimate of θ may be determined approximately by subtracting 1 unit from the maximum lod score and reading off the values of θ on either side of this value. The greater the maximum lod score and the narrower the 95% probability limits the more confidence there is in the estimated value of θ (Fig. 7.3). Very approximately 1 centiMorgan corresponds to 1 megabase (1 million base pairs), although there are exceptions to this, particularly near the gene for Duchenne muscular dystrophy, which will be discussed next.

These calculations are not difficult but can be tedious and therefore investigators often resort to using a computer program specifically designed for calculating lod scores (e.g. LIPED devised by Ott (1974) and LINKAGE by Lathrop and Lolouel (1984)).

PRENATAL DIAGNOSIS OF DUCHENNE MUSCULAR DYSTROPHY

This disorder is named after Guillaume Benjamin Amand Duchenne who, as an itinerant physician in Paris, described the condition in detail in 1868, though it had

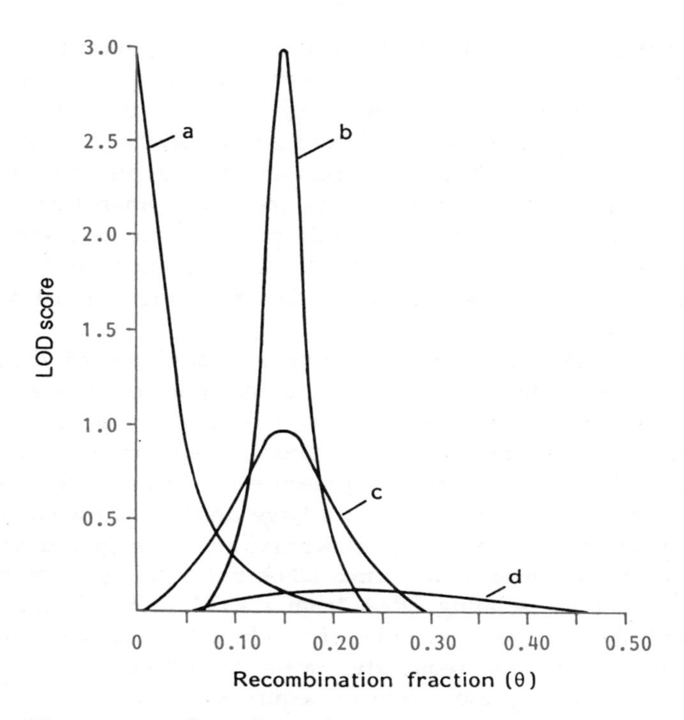

Figure 7.3 Diagram of lod scores plotted for various values of the recombination fraction (θ) indicating: (a) high probability (1000:1) of very close linkage (θ less than 0.01), (b) high probability (1000:1) that θ equals 0.15, (c) suggestion (10:1) of linkage, (d) no discernible linkage.

been described several years earlier by Edward Meryon of St Thomas' Hospital, London. The onset is usually before the age of 5 and is characterized by progressive muscle wasting and weakness which predominantly affects the proximal limb musculature and is always associated with enlargement of the calf muscles. Weakness gradually progresses, so that affected boys become wheelchair bound by their early teens and usually die in their late teens or early twenties. Being inherited as an X-linked recessive trait it only affects boys and is transmitted by female carriers. The gene product is dystrophin, a very large protein of molecular weight 427 kDa consisting of 3685 amino acids. Dystrophin is a rod-shaped molecule apparently joining actin fibres to the sarcolemmar membrane via a glycoprotein complex.

Over the years a great many clinical, biochemical, histological and electrophysiological tests have been devised for detecting female carriers of the disease. The single most reliable test has proved to be the serum creatine kinase (SCK) level which is raised in roughly 70% of carriers. In order to calculate the likelihood of a female relative of an affected boy being a carrier, not only is pedigree information taken into account but also her SCK level, the information being combined using Bayesian statistics. Prior to DNA techniques being introduced, the effect of basing risks on SCK data as well as pedigree data reduced considerably the proportion of women who fell into the intermediate risk range, most proving to be either at high risk (greater than one in 10) or at low risk (less than one in 20 of being carriers). Although measurement of SCK levels was valuable it was not without problems. There is considerable overlap in SCK levels in controls and carriers and because of this and the problems of X chromosome inactivation, it was never possible to reassure any woman that she was **not** a carrier. Both closely linked and intragenic polymorphic markers are used for indirect detection and gene analysis, mRNA and translation products are used for direct detection. It is often possible now to change a woman's risk to greater than 99% or less than 1%.

The locus for Duchenne muscular dystrophy is located towards the middle of the short arm of the X chromosome (Xp21) and the first gene-specific probes were isolated using material from a unique boy who had, in addition to Duchenne muscular dystrophy, mental retardation, chronic granulomatous disease, retinitis pigmentosa and Mcleod syndrome. The presence of a group of unrelated disorders like this is suggestive of the presence of a large deletion particularly as, being X-linked, the boy will have no other copies. Kunkel and colleagues devised a method for isolating sequences present in normal DNA but missing in the patient with a deletion. This involved isolating DNA from a 49XXXX,Y cell line and partially digesting it with the restriction enzyme *Mbo*I, which has only four base pairs in its recognition site and so cuts frequently. Partial digestion with a frequent cutter should ensure random representation of sequences each of which will have an overhanging sticky end produced by the *Mbo*I. DNA from the patient was randomly sheared. These segments of DNA will have random irregular ends. The two sets of fragments were mixed, with the randomly sheared sample from the patient in considerable excess, and heated to separate the strands and then cooled slowly together. Because of the presence of phenol in the buffer to speed the reaction the method is called the PERT (phenol enhanced reassociation technique) and the probes produced were called the PERT series. Three types of reassociated molecule will be produced: the majority will be from the patient's DNA and will have ragged ends, some duplexes will have formed with one strand from the patient and the other from the cell line; these will have an *Mbo*I overhang at one end only. However, for any DNA strands mapping to the deleted region there will only be copies from the cell line and they will be forced to reassociate with themselves. They will therefore have *Mbo*I sites at both ends and can be cloned into a suitable vector. Some of the PERT probes were found to be deleted in about 10% of boys suffering from Duchenne muscular dystrophy which made it very likely that they came from within the gene itself. At very much the same time Ron Worton and colleagues isolated sequences from the Duchenne locus using DNA from a female, with DMD, carrying a translocation between Xp21 and the ribosomal RNA genes on chromosome 21 (the XJ series).

Indirect analysis using RFLPs and linkage

As discussed earlier (page 72), DNA variations produce alterations in restriction sites which generate restriction fragment length polymorphisms (RFLPs). These appear to be randomly distributed throughout the genome and can be used as linkage markers for detecting genetic disease. The establishment of linkages between disease loci and various RFLPs has been a growth industry throughout the 1980s and has proved to be particularly valuable in a whole range of both autosomal recessive and X-linked disorders. However, there are in fact practical problems inherent in this approach, some of which are becoming less with time. Firstly, linkage must be close, otherwise there is the real possibility that crossing-over will occur between the DNA marker and the mutant gene in question. The likelihood of this occurring can be calculated and combined with other relevant genetic information and then translated into an overall risk figure. Though this can be helpful it is not ideal and may leave the risk in an intermediate area which makes decision making particularly difficult. In the case of DMD, as the original probes revealed deletions in affected boys, it was clear that the probes must come from within the gene. The first assumption was that if an intragenic probe is used the rate of recombination will be very low indeed, but in the case of DMD, family studies showed that a recombination rate of at least 5% should be used, even for probes within the gene. This was an early indication that the gene causing DMD is exceptionally large. Secondly, even when a close linkage or intragenic polymorphism is found by chance, relatives may not have the appropriate RFLP (so-called uninformative). Thus it is important to find an RFLP which is relatively common in the general population (at least one in ten) so that there is a reasonable chance of it occurring in a family being studied. The advent of variable number tandem repeats (VNTRs) has greatly alleviated this problem. Thirdly, the linkage phase in a family must first be established and this requires analysis of the parents and at least one child, or the grandparents or lateral relatives of the patients. Again the increasing use of the PCR, in which small amounts or very poor DNA can be used as a template, has helped here because sometimes DNA from a dead child can be extracted from a Guthrie spot or some other pathological sample. Finally, in serious X-linked recessive disorders there is the added problem that when there is only one affected male in a family this may well represent a new mutation and in this situation linkage may not be helpful or may even be misleading (page 139).

In general in calculating genetic risks, and particularly in X-linked disorders, **Bayesian statistics** are often used, so named after the Reverend Thomas Bayes whose theorem was published posthumously in 1763. Details of the application of Bayes' theorem to calculating genetic risks can be found in several texts (e.g. Emery, 1986). Essentially, the method consists of combining **prior** risks, i.e. risks based on information about the woman's antecedents, with **conditional** risks based on the results of various biochemical tests and the number of normal and affected sons she may have. However, for the sake of simplicity this refinement will be ignored in the following discussion and in any event has become less important in genetic counselling as opportunities to study the DNA changes in the gene itself have become available.

The application of linkage with an RFLP in carrier detection and prenatal diagnosis in Duchenne muscular dystrophy can be illustrated with a few examples. In Fig. 7.4(a) the mother must be a carrier because she has two affected sons. If linkage with an RFLP were very close, and since both sons possess the same polymorphic allele, then it is possible to predict the status of any future son (individual 5 would be normal) or daughter (individual 6 would **not** be a carrier).

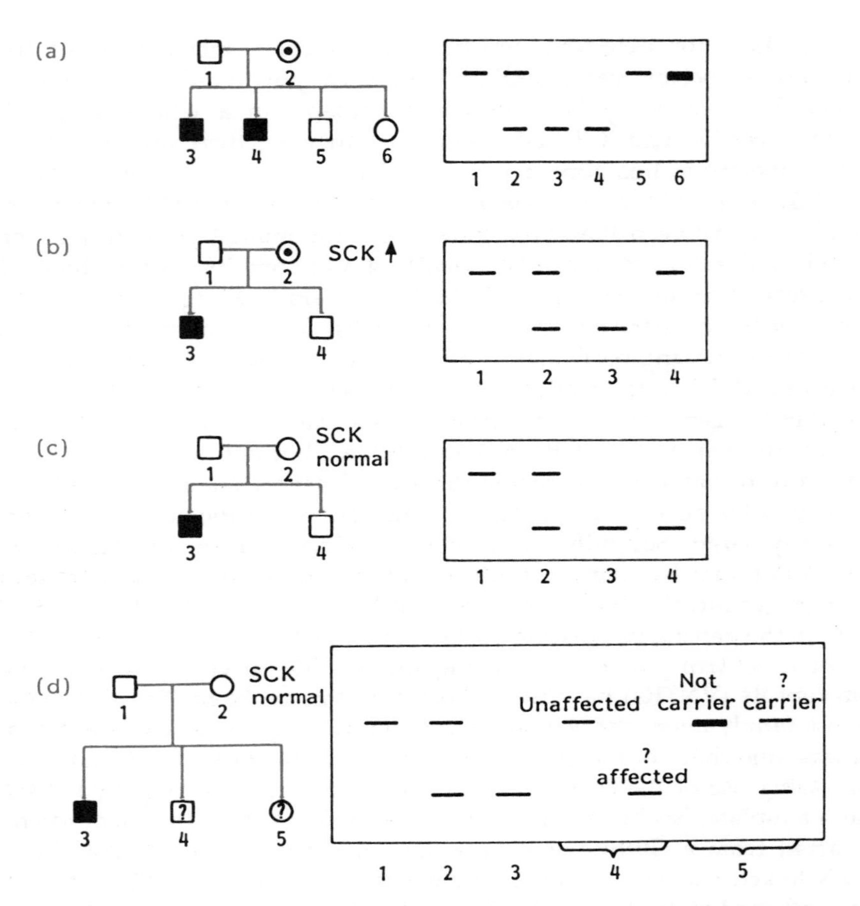

Figure 7.4 Application of linkage with a restriction fragment length polymorphism in carrier detection and prenatal diagnosis in X-linked Duchenne muscular dystrophy. DNA fragments on gel electrophoresis are diagrammed on the right.

But if the RFLP and the disease locus were only loosely linked, then crossing-over between the disease locus and a polymorphic locus could occur, and the chance of the son being affected or the daughter being a carrier in this example would then be equal to the chance of recombination occurring between the two loci. Clearly the closer the loci the more helpful would be the information from a DNA marker. The prior risk of a daughter in Fig. 7.4(a) being a carrier is 50%, but if she did not inherit the polymorphic allele in coupling with the disease locus in her mother, her chance of being a carrier would, as we have seen, be reduced to the probability of recombination. Recombination around the dystrophin locus is remarkably high. In fact, recombination between probes at the extreme 5' and 3' ends is 12%. Therefore, it is unsafe to rely on linkage using any one polymorphism, unless the position of the mutation within the gene in a particular family is known, and polymorphisms based on CA repeats have now been developed covering the whole gene (Fig. 7.5). It should be noted that an affected relative must be available in order to establish

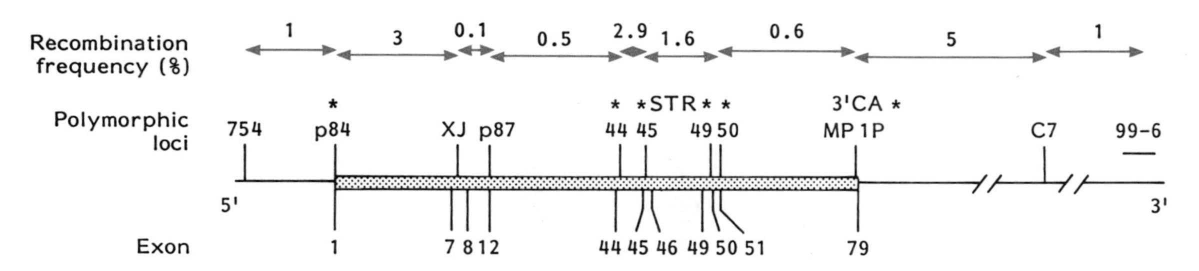

Figure 7.5 Recombination frequencies in the dystrophin gene. * = (CA) repeat.

the linkage phase in the family being studied, or at least some stored DNA. DNA banking, particularly from individuals at risk of dying, has become an important activity for genetic centres.

With many serious X-linked diseases there is yet another problem, namely that there is only one affected boy in the family. Genetic theory predicts that in an X-linked disorder where affected males do not reproduce, one-third of all cases will be new mutations. The problem is to know whether an **isolated (sporadic)** case in a family is in fact a new mutation or whether the mother is a carrier, in which case other female relatives would then be at risk. If the mother of an isolated case has a sufficiently raised SCK level, then she is most likely to be a carrier, and data on linked RFLP can therefore be useful for prenatal diagnosis and determining the carrier status of any future daughters she may have (Fig. 7.4(b)). If, however, her SCK level is within the normal range and an affected son and a normal son have both inherited the same allele it would seem likely that the affected son is a new mutation. Even this assumption has now been understood to be too simple as many examples have been found of **germ-line mosaicism**. Bakker and colleagues in Leiden analysed a family in which a deletion of the probe PERT 87-15 had been transmitted to two sons by a woman who did not have the deletion herself in DNA extracted from blood. She must, therefore, have been a mosaic with the deletion having arisen in her germ cells (Fig. 7.6). Darras and Francke described a woman from a four generation family, where two of her five daughters had sons affected with DMD, who carried an intragenic deletion even though she (the grandmother) did not have this deletion. Study of the haplotypes showed that the daughters had inherited the mutant chromosome from their unaffected father. Therefore, germ-line mosaicism is found in both males and females. In practical terms, it means that even if a boy with DMD has been shown to be a fresh mutation, there is still a chance of his mother having another affected boy and she should consider prenatal diagnosis. This example illustrates well why direct detection of the gene mutation is so superior to the indirect methods, but the following section explains why even with a well characterized gene such as dystrophin this is not problem free.

Direct detection and the dystrophin gene

The complete dystrophin cDNA was isolated by Koenig and colleagues in 1987. The dystrophin gene has a remarkable organization. The mRNA is 14 kb long but the coding sequences are divided into 79 exons extending over 2.4 megabases. This

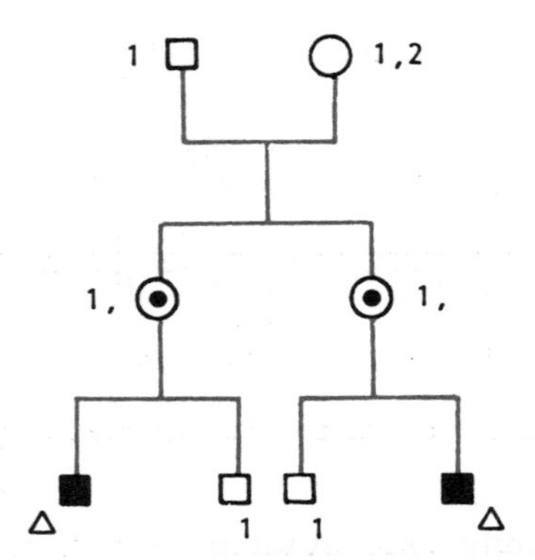

Figure 7.6 Pedigree showing germ-line mosaicism in Duchenne muscular dystrophy. RFLP results are given for the probe PERT 87-15. The grandmother is heterozygous and, therefore, must have two alleles and not carry a deletion. However, both her daughters have PERT 87-15 deleted as revealed in their affected sons. Grandmother is a germ-line mosaic for the deletion.

means that the coding regions make up less than 1% of the total gene. Some of the introns are very long, intron 44 is 160–180 kb long. It would clearly be a major undertaking to sequence the entire coding region in every affected boy and fortunately this has not proved necessary because a high proportion of all mutations have been found to be due to deletions of one or more exons. About 70% of affected boys have a deletion which can be detected using the Southern blot method. DNA is digested with a suitable restriction enzyme, such as *Bgl*II, which, because the gene has such a high proportion of intervening sequence compared to coding sequence, will cut on average once within each intron. Probing with radioactively labelled dystrophin cDNA will then reveal any 'missing' or deleted exons. If a full length cDNA were used as a probe the ensuing result would be too difficult to interpret because with so many exons there would be too complex a pattern. Shorter regions of cDNA are used. This is made easier by the observation that the deletions tend to cluster in certain regions of the gene, particularly around the very long exon 44 (Fig. 7.7).

Again the polymerase chain reaction has simplified the analysis of mutant dystrophin genes. As the exons are so short, often only 100–200 base pairs long, they can readily be amplified using suitable primers from within the intervening sequences surrounding each exon. The presence or absence (deletion) of a product can be visualized by running the products on a gel and staining with ethidium bromide, a method which is quick and involves no radioactivity. The practical problem is the number of exons to be amplified. The method has been simplified by Chamberlain. Experience in many patients has shown that amplification of 18 exons, in two batches, will detect 98% of all deletions (see Fig. 7.7) and he has designed sets of primers which can be used together in one PCR reaction to analyse several exons at a time (so-called multiplex PCR). This requires some skill in planning as each primer must work at a similar annealing temperature, there must be no chance of pairing between any primer pairs and the size of each exon product must be sufficiently different that they can be separated on a gel (Fig. 7.8).

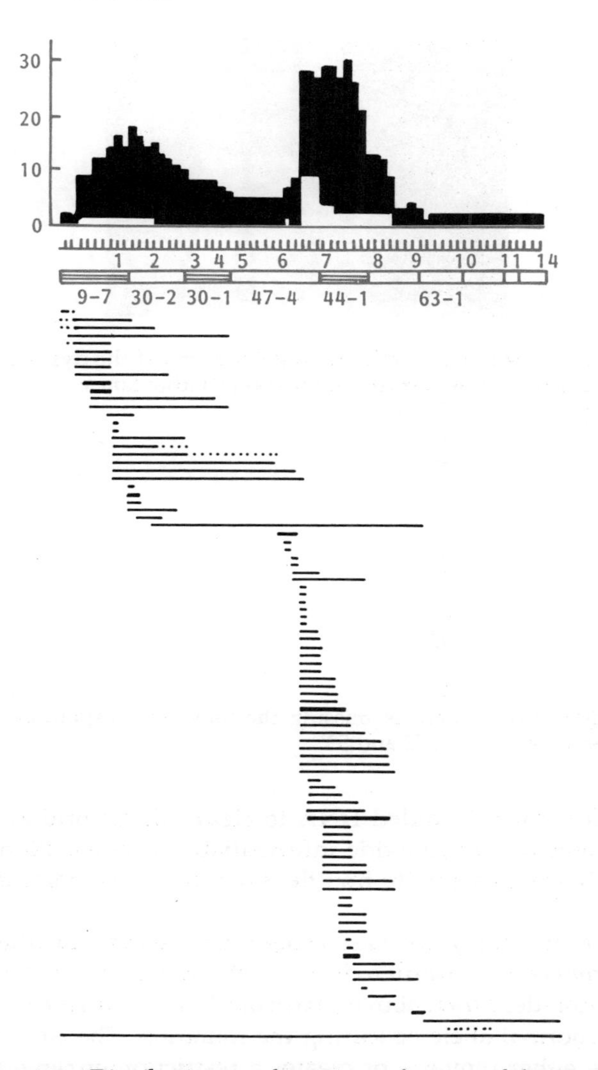

Figure 7.7 Distribution and extent of gene deletions and duplications in Duchenne (solid) and Becker (open) muscular dystrophies. (From Bakker, 1987, with permission.)

This still leaves the problem of finding base changes in the 30% of cases where there is no deletion. It would be much easier to work directly with the mRNA as no effort would be wasted on non-coding sequences. However, as the major site of dystrophin mRNA production is in muscle it is not easy to obtain a suitable source. An approach to this problem was suggested by the finding that there is a very small amount of illegitimate, or ectopic, transcription of most genes including dystrophin in cell types which are not normally expressing the gene. Dystrophin cDNA can be produced from lymphocytes using RT-PCR and nested primers (see page 52). The site of mutation is then found by mixing the dystrophin transcript with normal cDNA and using chemicals (hydroxylamine and osmium tetroxide) which

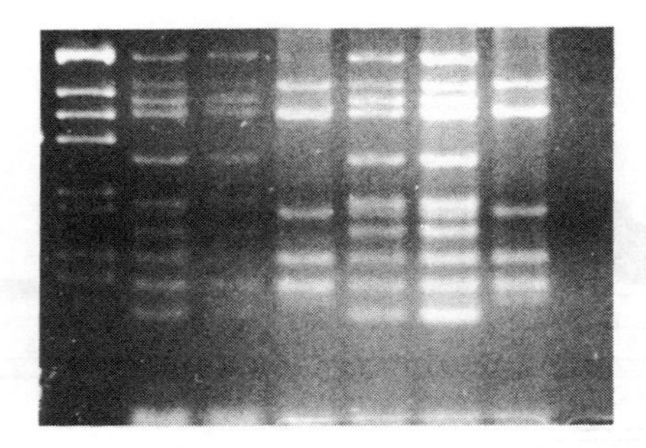

Figure 7.8 Multiplex PCR analysis of exons of the dystrophin gene. Lane 1: molecular weight markers. Other lanes:

Base pairs	Exon
506	48
426	44
388	51
357	43
271	50
212	53
202	6
181	47
155	42
139	60
113	52

In lane 4 the patient is missing the bands corresponding to exons 47, 48, 50, 51, 52 and 53.

chemically modify single stranded DNA to cleave the hybrid at any mismatched sequences. This method will provide information on the position of the sequence alteration and DNA sequencing to find the exact change can be confined to a short region (Fig. 7.9).

As the purpose of finding the base change is to search for the same change in other family members, it is useful to have a method which is less involved than the ectopic transcription described above, particularly in the analysis of carrier women who will have a normal allele obscuring the mutation. The easiest situation arises when the change either removes or creates a restriction enzyme site. The product from all family members can be digested with the restriction enzyme and scored for presence or absence of the site. However, this is frequently not the case, but the problem can be got round by an ingenious method called ACRS or artificial creation of restriction sites. In the PCR, specific binding of the nucleotides used for amplification is very important but a single base mismatch will not reduce binding significantly. Providing that the mismatch is not at the 3' end base where extension occurs, amplification will still take place. It is possible to design an alteration in the sequence such that a restriction enzyme site which does not exist naturally is created. Its presence or absence in the final product will be determined by the mutation of genomic DNA just beyond the start of amplification. An example is illustrated in Fig. 7.10.

As information about point mutations in the dystrophin gene accumulated it became apparent that quite a lot of polymorphisms would be tolerated in the

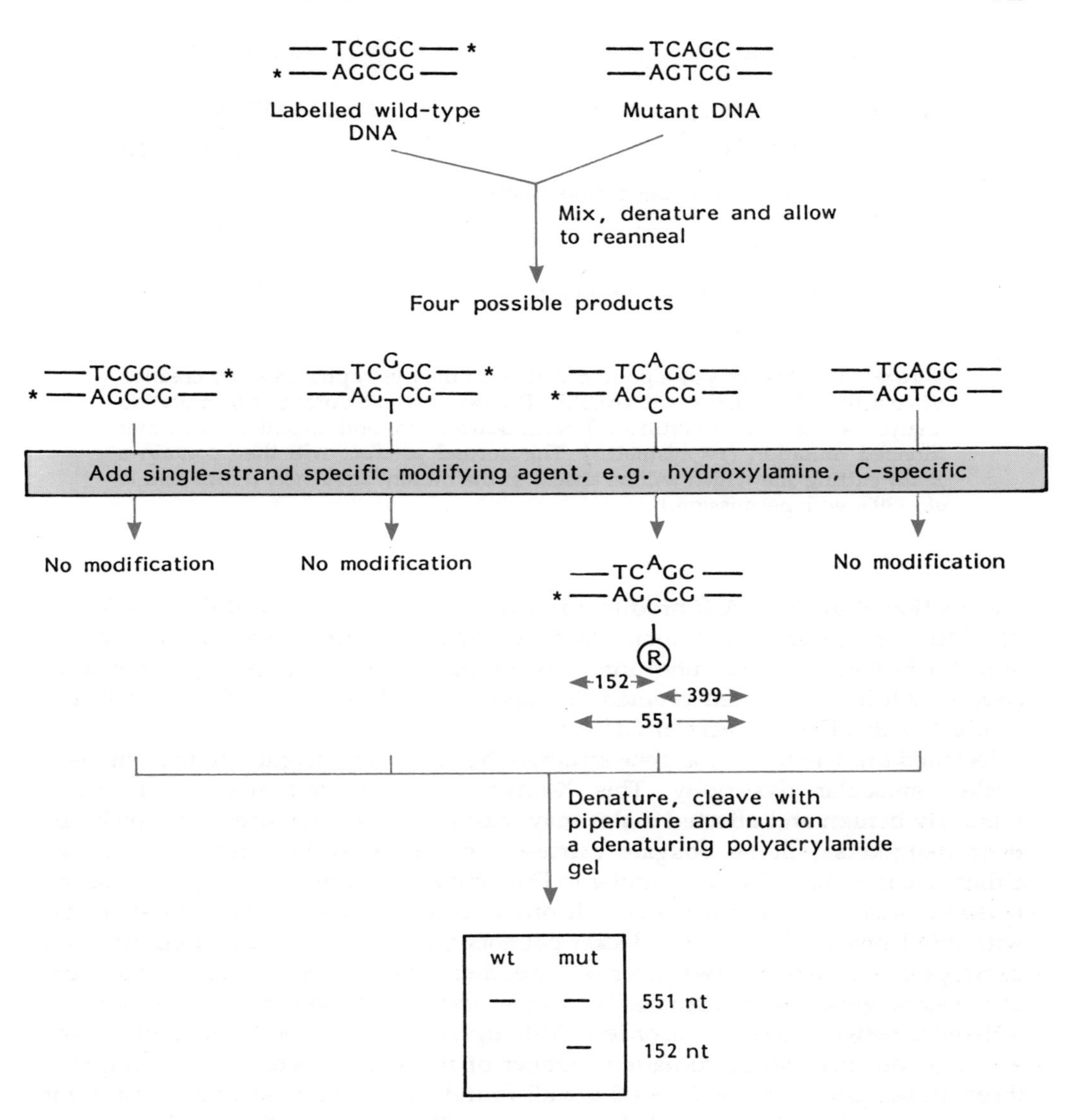

Figure 7.9 **Identification of mismatches by chemical cleavage. * = label, ^{32}P or non-radioactive.**

coding sequence without causing any pathological changes. Most of the disease-causing point mutations were in fact alterations which led to premature termination codons. An ingenious method for detecting these has been devised and is shown in Fig. 7.11. Obviously the best way to search for a termination mutant is to test the product of translation itself. This can be achieved by the following method. The primers for PCR are again altered but this time by synthesizing a stretch of DNA which is a strong bacterial transcription promoter onto the 5' end. This will not bind to genomic DNA but, being at the 5' end, will not interfere with

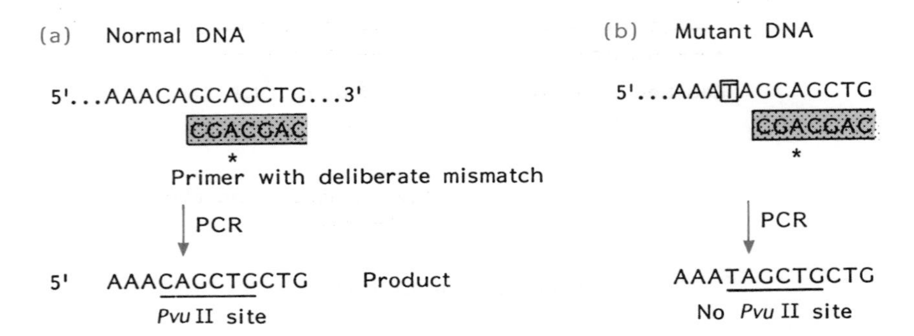

Figure 7.10 Analysis of a point mutation in the dystrophin gene by creation of an artificial restriction site (ACRS). The two possible outcomes of mutation analysis are shown. An artificial T is introduced into both products by primer directed mutation (* = mismatch). The normal product will then contain a *Pvu*II cutting site which will be absent in the mutant sequence. (From Yau *et al.*, 1993, with permission.)

the reaction at all. The PCR product can then be transcribed into RNA by addition of T7 RNA polymerase and ribonucleotides and this *in vitro* transcript translated *in vitro* (in fact in the same tube) into protein. Any changes containing termination codons will be terminated prematurely and this will be observed as a truncated product on an SDS–polyacrylamide gel.

Detailed knowledge of the gene structure has given an insight into the origins of Becker muscular dystrophy. This X-linked form of muscular dystrophy is relatively benign and affected males may marry and have children (in which case their daughters will be obligate carriers) and often survive into middle age. Otherwise it is clinically very similar to Duchenne muscular dystrophy and for this reason it was thought that the two disorders could be allelic. The earliest studies with PERT probes showed that Becker patients also frequently had deletions of the dystrophin gene and the two disorders are, therefore, indeed caused by mutations in the same gene. Further studies led to the 'frameshift hypothesis' to explain the difference between the two disorders. Although the genetic code is read in triplets, exons do not necessarily contain a number of nucleotides which is a multiple of three. In fact $3n$, $3n+1$ and $3n+2$ are all found. As we have seen the dystrophin gene is prone to deletions and the outcome will depend on the number of base pairs in the deleted exons. If an exact multiple of three is deleted there will be no frameshift (see page 27) and mature, albeit truncated, protein will be produced, giving rise to Becker muscular dystrophy. If the exons deleted do not remove an exact multiple of three base pairs then a frameshift will be produced often leading to the creation of a premature termination codon and the more severe Duchenne muscular dystrophy.

It is emerging that many isoforms of dystrophin exist resulting from the use of additional promoters or arising by alternative splicing. As a result specific mutations can give rise to differing disease presentations, for example a deletion of a muscle-specific promoter has been observed in a family with severe X-linked dilated cardiomyopathy but without skeletal muscle involvement.

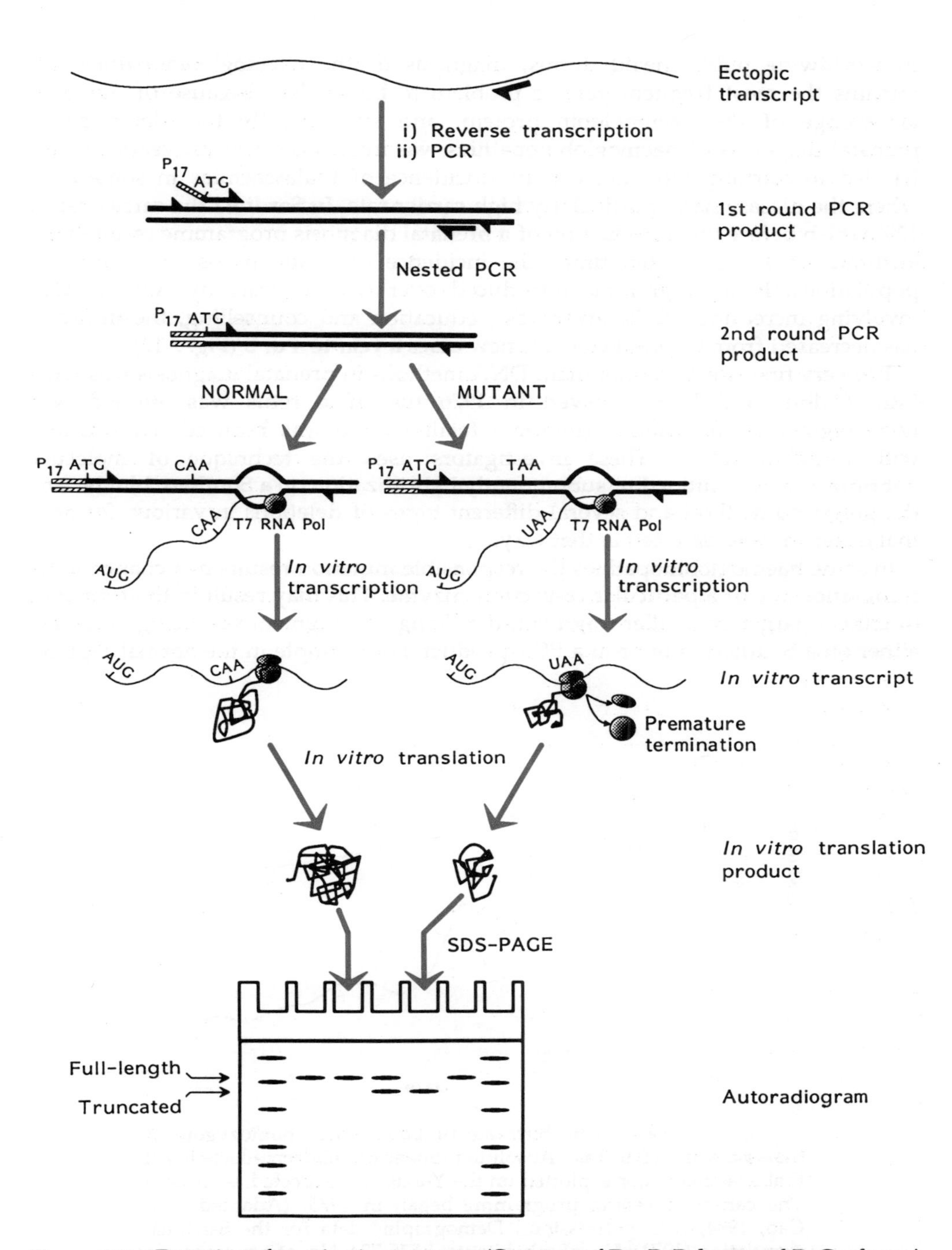

Figure 7.11 Detection of termination mutants. (Courtesy of Drs R. Roberts and R. Gardener.)

PRENATAL DIAGNOSIS OF THE HAEMOGLOBINOPATHIES

In worldwide public health terms, diagnosis of the haemoglobinopathies still remains the most frequent genetic problem to be tackled. Because of our early knowledge of the haemoglobin protein, and subsequently the globin genes, prenatal diagnosis of haemoglobinopathies was introduced several years ago and has led to startling reductions in the incidence of thalassaemias in some areas where the disease has a particularly high carrier rate. In Sardinia the carrier rate is 12% which before the introduction of a prenatal diagnosis programme resulted in a birthrate of 1:250, or ten times the incidence of cystic fibrosis in Caucasian populations. In a programme introduced over twenty years by Antonio Cao, involving increasing public awareness, education and counselling, the incidence has decreased from the predicted 120 new cases a year to 4 or 5 (Fig. 7.12).

The very first use of recombinant DNA methods in prenatal diagnosis was when Kan, Golbus and Dozy showed in 1976 that if a fetus was affected with homozygous α-thalassaemia (hydrops fetalis) there was reduced hybridization with α-globin cDNA. These investigators used the technique of molecular hybridization in solution but subsequently hybridization on a Southern blot became the preferred method and several different types of deletions in various forms of thalassaemia were detected in that way.

In some haemoglobinopathies the responsible mutation results in a change in the recognition site of a particular restriction enzyme. This may result in the formation of either a larger or smaller fragment, the change in fragment size being detectable either on a Southern blot or in a PCR product. For example in the normal β-globin

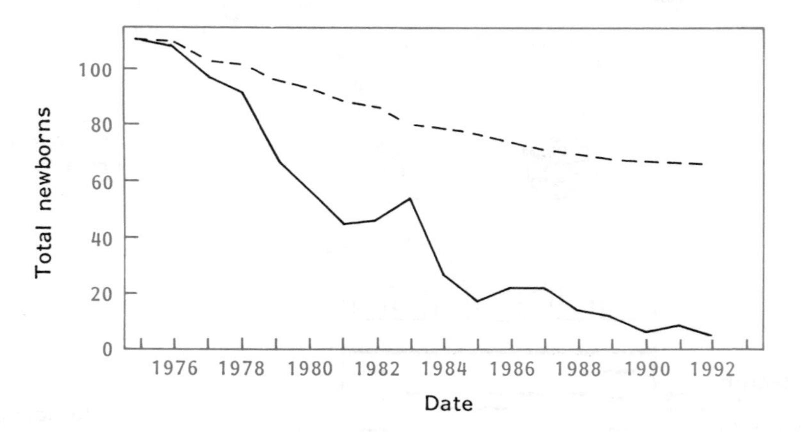

Figure 7.12 Fall in the birthrate of babies with homozygous β-thalassaemia in Sardinia. Absolute number of children affected with thalassaemia major is plotted on the Y-axis. ---, expected; —, found. The carrier screening programme began in 1975. (Adapted from Cao, 1994, with permission.) Demographic data for the Sardinian population (1974): No. of inhabitants: 1 535 724. No. of newborns per year: 29 881. Birth rate of thalassaemia major: 1:250. Total no. of new cases of thalassaemia major per year: 120

*Dde*I site

βA	ACT	CCT	GAC	GAG	AAG	TCT	GCC
codon	4	5	6	7	8	9	10
βS	ACT	CCT	GTG*	GAG	AAG	TCT	GCC

Figure 7.13 Detection of HbS by restriction enzyme site changes. The recognition site for *Dde*I is CTNAG.

gene there is an *Eco*RI site (G*AATTC) in the region which encodes amino acids 121 and 122. However, in HbO_{Arab} the responsible mutation (a nucleotide substitution of G to A) occurs at this point and eliminates the *Eco*RI site. The result is that when normal genomic DNA is digested with *Eco*RI and hybridized with a β-globin gene probe two DNA fragments of 3.2 kb and 5.2 kb are produced, but in HbO_{Arab} only a single fragment of 8.4 kb is produced. Another most important example is the detection of sickle cell haemoglobin, HbS, which results from an A to T change in codon 6 of the β-globin chain causing the substitution of valine for glutamic acid. This results in the change of the restriction site for either *Dde*I (or the more expensive *Mst*II) (Fig. 7.13).

Considerable ingenuity has been exercised in choosing suitable enzymes for detecting haemoglobinopathies in this way. Unfortunately only a few disorders have proved amenable to this approach because most mutations do not affect any known restriction site or the enzyme may be rare and expensive. An alternative

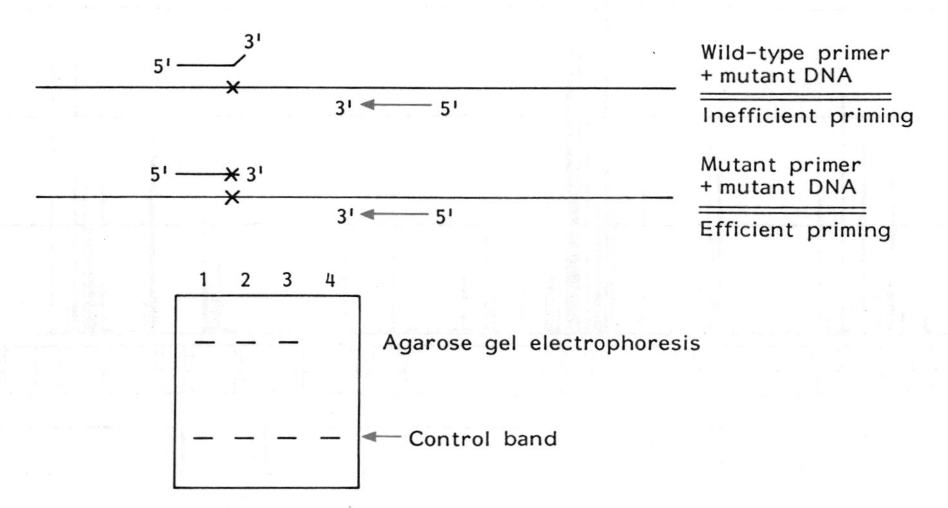

Figure 7.14 Amplification refractory mutation system (ARMS). * = mutant sequence. Lane 1: Sample from heterozygote plus mutant primer. Lane 2: Sample from heterozygote plus wild-type primer. Lane 3: Sample from patient homozygous for mutation plus mutant primer. Lane 4: Sample from patient homozygous for mutation plus normal primer.

strategy, again using the convenience of the PCR reaction, has been devised called ARMS for amplification refractory mutation system. We have already discussed that even oligonucleotides with one or two bases mismatched will bind relatively specifically to genomic DNA, but their efficiency in acting as primers for extension will be drastically reduced if there is a mismatch at the 3' end. The ARMS technique uses oligonucleotides with their 3' ends corresponding to a known proposed site of mutation and compares the efficiency in priming of the normal and wild-type sequence (Fig. 7.14). In general another unrelated pair of primers, with a different size product, is included to act as a control that the PCR conditions were suitable. The ARMS technique is used to analyse thalassaemia samples from India where five mutations account for 90% of cases. One of the common mutations is a 619 bp deletion towards the 3' end of the β-globin gene. A pair of primers flanking this region is used as the control, the product will also act as a check for the presence of the deletion. The other four mutations are a G to C change at the fifth position of the first intron (IVSI), an additional G after position 8 leading to a frameshift, a G to T at the invariant splice site of IVSI and a 4 base pair loss leading to a frameshift after position 41. ARMS primers have been designed to all of these. If none come up positive the next five most common mutations are searched for. It is potentially easier to transfer technology based on the PCR and not using radioactivity to less developed countries.

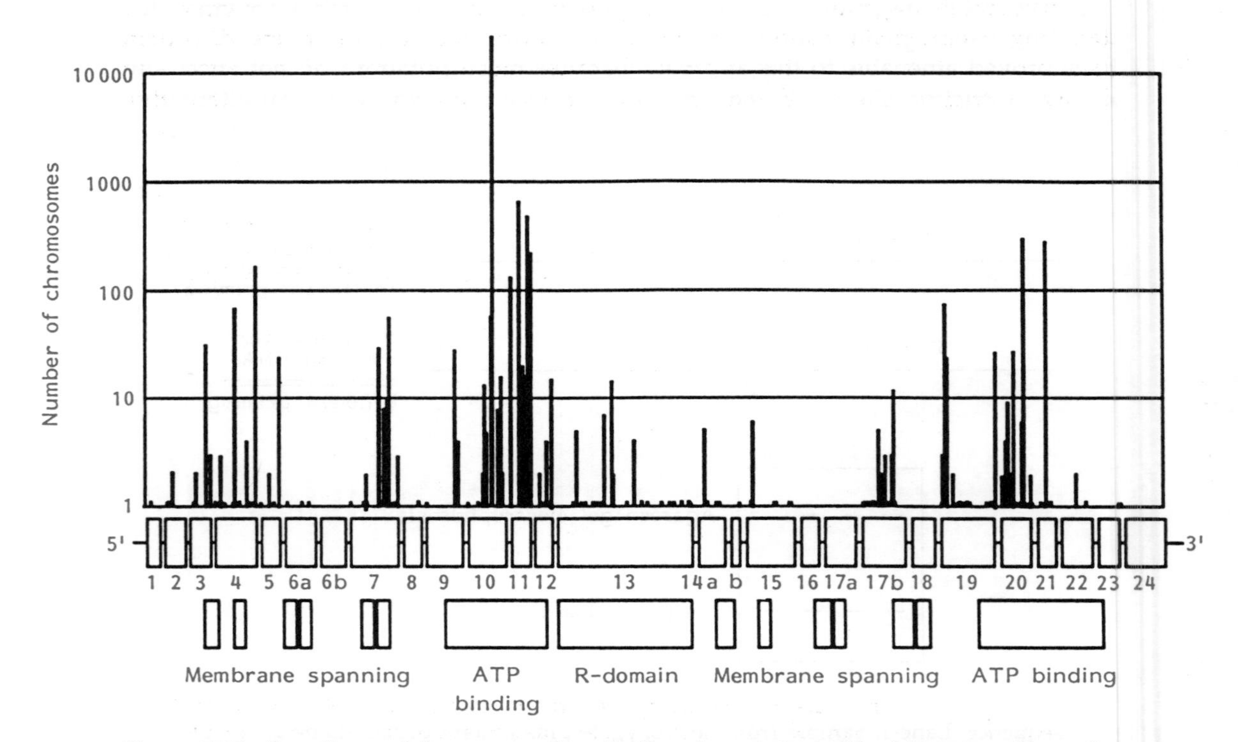

Figure 7.15 Location and frequency of cystic fibrosis mutations. The histogram shows the number of cases of each putative mutation with respect to its location in the CFTR gene. (From Tsui, 1992, with permission.)

CYSTIC FIBROSIS

Whereas in Mediterranean regions and the Far East, haemoglobinopathies produce the major genetic burden, amongst Caucasians the highest carrier rate, about 1 in 20, is for the gene causing cystic fibrosis. This is a severe autosomal recessive disorder in which altered chloride ion transport leads to the accumulation of viscous mucus resulting in chronic obstructive lung disease and pancreatic insufficiency. Although elevated sweat electrolyte concentration was known to be part of the disease the nature of the biochemical defect was unknown until, after an enormous international collaborative effort, the gene was isolated in 1989 and shown to be a cyclic AMP induced chloride channel. The gene contains 27 exons and the mRNA is 6129 base pairs long. The protein includes two hydrophobic transmembrane domains, two nucleotide-binding sites and a highly charged regulatory domain (Fig. 7.15). It was named cystic fibrosis transmembrane conductance regulator (CFTR). Prenatal diagnosis using linked probes was available from 1986 onwards for families with a child already affected with cystic fibrosis. In complete contrast to X-linked disorders, a single major mutation, a 3 base pair deletion resulting in the loss of phenylalanine at amino acid position 508 (within a nucleotide binding domain), has been found in Northern European populations. In Britain this accounts for about 75% of all mutant chromosomes. The mutation can readily be analysed by PCR amplification using primers on either side of mutation. In carriers of the Δ508 deletion two bands will be seen differing in size by 3 base pairs (Fig. 7.16). An additional aid to detection is the production of a band migrating slower in the gel caused by heteroduplex molecules containing one full length strand and one deleted strand. As these mismatched molecules will have a slightly irregular structure they will be slowed down as they pass through the pores of the gel. However, after the finding of this major mutation the picture is not so simple. In a worldwide study by a consortium containing 90 laboratories from 26 countries, only four other mutations have been found with frequencies of greater than 2% adding up to 10% additional mutations in total (Table 7.2). After that there are over two hundred other mutations, many of them having been found only once. The distribution of all the mutations within the CFTR molecule is shown in

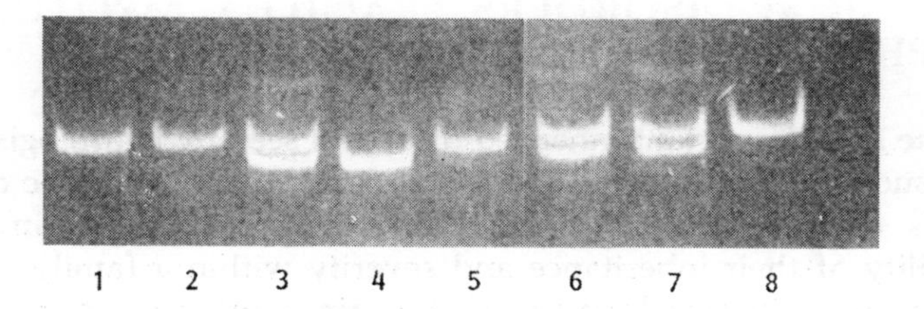

Figure 7.16 PCR analysis of the 3 base pair deletion causing most cases of cystic fibrosis. The lower, faster migrating, band is smaller by 3 base pairs. The samples are lane 1: normal; lane 2: normal; lane 3: carrier of mutation; lane 4: homozygote for mutation, i.e. patient with cystic fibrosis; lane 5: normal; lane 6: carrier; lane 7; carrier; lane 8: normal.

Table 7.2 Most frequent cystic fibrosis mutations

Mutation	Relative frequency	Consequence
Δ508 Deletion 3 bp 1652–1655 in exon 10	67.2	Deletion of Phe at codon 508
G542X G → T at nt 1756 in exon 11	3.4	Gly → stop at codon 542
G551D G → A at nt 1784 in exon 11	2.4	Gly → Asp at codon 551
W1282X G → A at nt 3978 in exon 20	2.1	Trp → stop at codon 1282
3905 Insertion T after nt 3905 exon 20	2.1	Frameshift

NB. Only 22% of the Ashkenazi Jewish population in Jerusalem carry Δ508 but the frequency for W128X is 60%.
Data from the CF genetic analysis consortium.

Fig. 7.15. Because of the high carrier level in the general population, screening for carriers would be desirable but using DNA based technology it will never be possible to screen in such a way as to ensure that an individual is definitely **not** a carrier. On the other hand, finding a Δ508 mutation shows that the person is definitely a carrier. A method which could detect at least the most commonly occurring mutation on a large scale screening basis has been devised based on a reversed form of ASO analysis (see page 103) and called reversed dot blot analysis. In this oligonucleotides are synthesized representing each of the mutations to be screened and a long homopolymer tail of poly d(T) added using the enzyme terminal transferase. These molecules are covalently attached, by cross-linking with ultraviolet light, to a nylon membrane in dots alongside their normal equivalent. A multiplex PCR (see page 142) is then carried out amplifying genomic DNA from the subject for each of the regions to be screened. The primers used in the PCR have biotin bound at the 5' end. This mix is then denatured and hybridized to the test strip. Hybridization is detected non-radioactively by binding of streptavidin–horseradish–peroxidase to the biotinylated DNA followed by simple colorimetric reaction.

NEUROLOGICAL DISORDERS, FRAGILE X, MYOTONIC DYSTROPHY AND HUNTINGTON'S CHOREA

Several of the most important inherited disorders seen by neurologists and clinical geneticists, such as fragile X-related mental retardation, myotonic dystrophy and Huntington's disease, had for many years posed problems in the apparent unpredictability of their inheritance and severity within a family. Although they are clinically very different because of this similarity and, as we now know, a similarity in the underlying mechanism, it is convenient to consider them together. In general, the disorders appeared to become more severe in subsequent generations. In the case of fragile X mental retardation (FRAXA) this manifested itself by the existence of males who were themselves quite normal but whose grandsons by their daughters had the disease. Yet they themselves must have been carriers because other relatives on their mother's side were affected and showed the

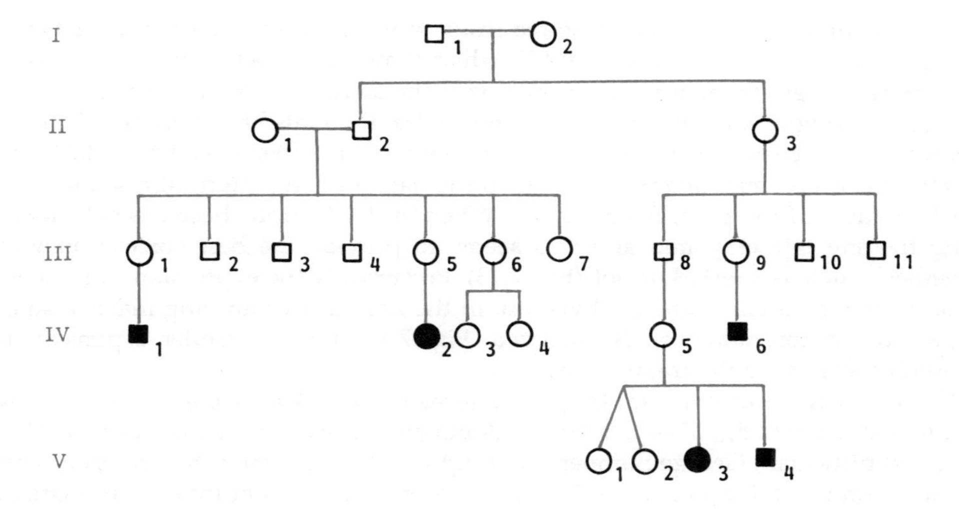

Figure 7.17 Fragile X mental retardation pedigree showing normal transmitting males. The gene has been transmitted through the normal transmitting males II-2 and III-8. The premutation has expanded to the full mutation in females III-1, III-5, III-9 and IV-5.

typical pattern of X inheritance (Fig. 7.17). In the case of myotonic dystrophy the severity of the disorder usually increases over three or four generations. A grandparent may be so mildly affected, with only cataracts, that they were never diagnosed as having myotonic dystrophy until other family members were affected. In further generations other symptoms may appear until the full blown severe disorder is seen with myotonia (difficulty in relaxing muscle after voluntary contraction), progressive muscle weakness and a variety of other manifestations including premature baldness, presenile cataracts and various endocrine abnormalities. It is therefore a serious disease and because in the milder cases onset is usually in adult life individuals may well have had a family before they become aware of the symptoms. In view of the tendency of the disease to become more severe in subsequent generations accurate presymptomatic diagnosis is of great importance. This phenomenon is called 'anticipation' but its very existence was called into doubt because some clinicians thought it could be a result of ascertainment bias. In other words, families in which the current generation are severely affected are likely to be the ones coming to the clinicians' attention. Isolation of the genes for these, and a number of other late onset neurological conditions, has shown that the phenomenon does indeed exist and has provided an explanation at the molecular level, although both the detailed mechanisms and the normal functions of the genes remains a mystery.

We have already referred to the presence in the human genome of runs of simple sequence repeats used as genetic markers and for fingerprinting. All the examples of genetic instability referred to above result from the presence of runs of triplet repeats, rich in the nucleotides C and G, which jump in size between generations; in general the longer the repeats the more severe the disease or the earlier the onset. In the case of fragile X mental retardation normal individuals have between 7 and 50 copies of the repeat but in affected males the copy number goes up to 2000 and more. This is accompanied by a change in the methylation pattern of a sequence at the beginning of the gene, the promoter. When the CCG triplet becomes sufficiently long the site is methylated and this seems to provide the best correlation with whether a male is affected or not (Fig. 7.18). Presumably the expression of the gene product is regulated by the methylation. In the normal transmitting males a small expansion, or premutation, is observed. Fig. 7.18 shows a triplet expansion in individuals carrying the fragile X mutation.

Of all genetic disorders, Huntington's chorea must rank as being one of the most tragic and distressing. This autosomal dominant disorder is named after a New England physician, George Sumner Huntington, who described the condition with great accuracy and detail in 1872 in his one and only publication. Huntington's chorea is characterized by progressive dementia and paralysis associated with involuntary choreic movements. Sufferers usually succumb about 10 to 15 years after the onset of symptoms, which is usually in the third or fourth decades of life. This means that many will have completed their families before they develop symptoms and realize they carry the mutant gene.

The disorder occurs throughout the world. New mutations are extremely rare and most cases occur in very large families in which the disorder can be traced back

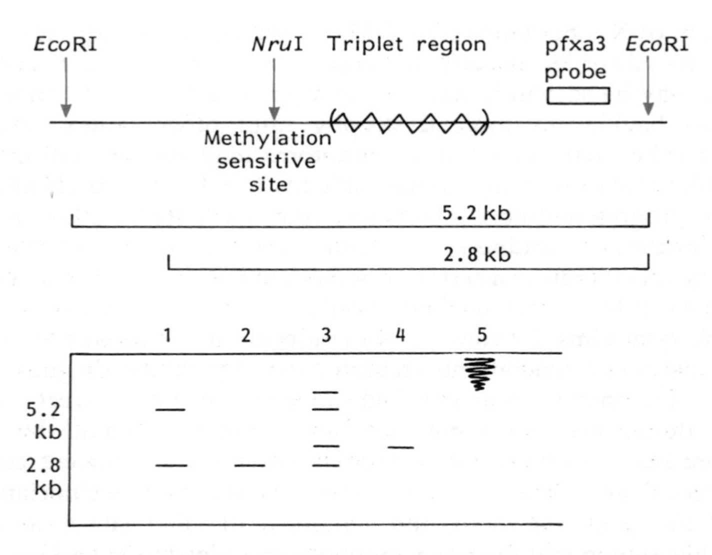

Figure 7.18 Detection of fragile X. Track 1: Normal female. One X is methylated. Track 2: Normal male. Track 3: Female with small premutation. Track 4: Normal transmitting male. Track 5: Affected male.

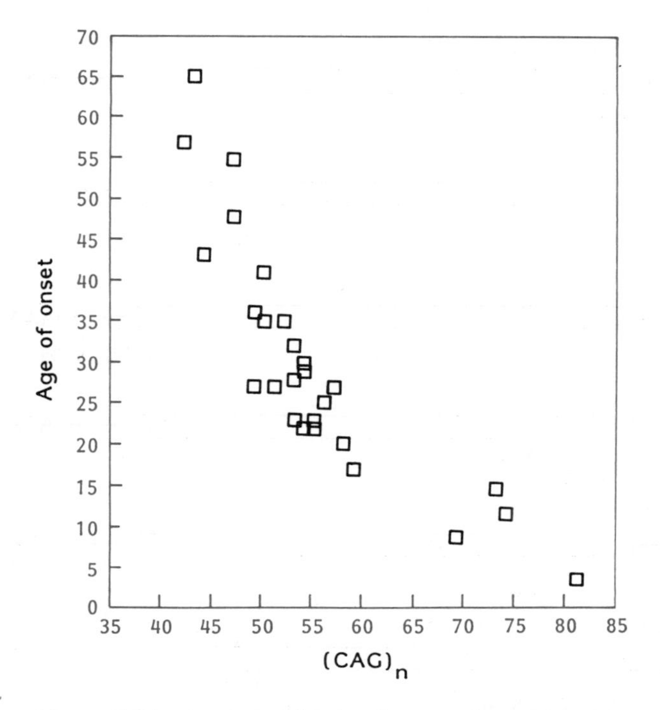

Figure 7.19 A scatter plot for the age of onset in years versus the number of the $(CAG)_n$ repeat units demonstrating the correlation between the age of onset and the size of the expansion in spinal cerebellar ataxia type I. (From Orr *et al.*, 1993, with permission.)

over many generations. In fact, many affected families in the United States can be traced to two brothers and their families who came from Bures in England and emigrated to America in 1630.

In Huntington's disease there is a general correlation between the number of repeats and the age of onset; the longer the triplet expansion the earlier the symptoms start. These data have been rather difficult to collect in Huntington's disease because the late onset and rapid progress of the disease means there are relatively few multi-generational families available for study. A further neurological disorder showing the same anticipation, spinal cerebellar ataxia type I, shows quite a clear relationship between repeat length and age of onset (Fig. 7.19).

The disorders in which this strange mechanism have been observed are summarized in Fig. 7.20. Fragile X mental retardation was named as such because of the observation of a fragile site or break in the chromosome in cytogenetic preparations made in a particular way. Although these fragile sites are quite common in human chromosomes they are rarely associated with any adverse effect. An exception is another X-linked fragile site FRAXE also caused by a triplet expansion.

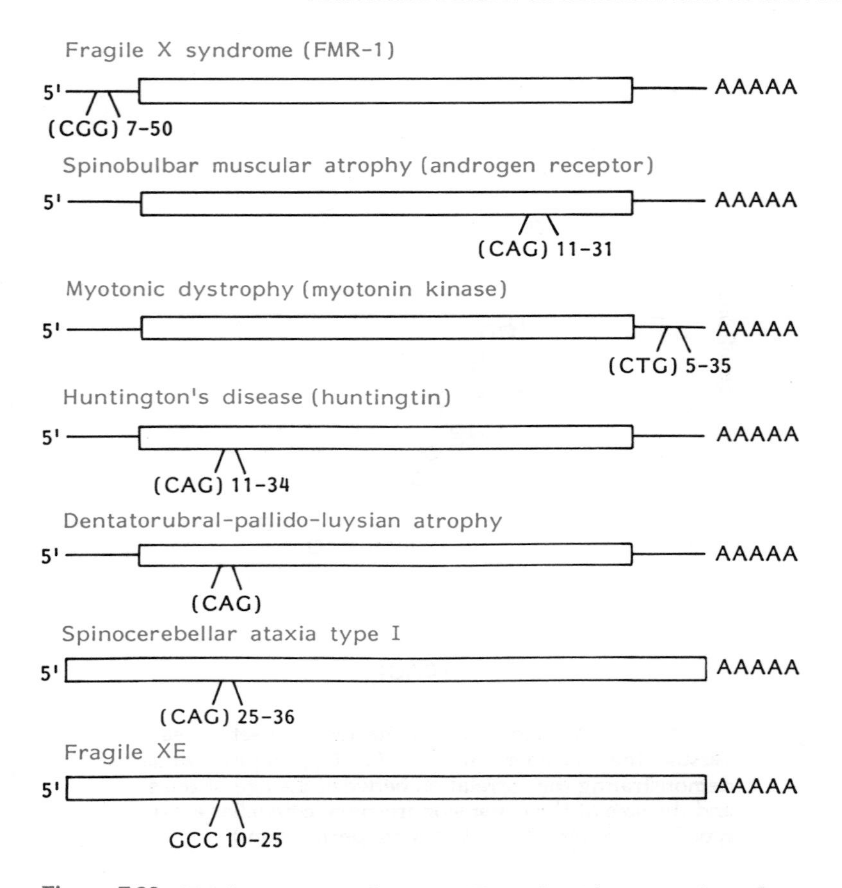

Figure 7.20 Triplet repeats in humans. Trinucleotide repeats have been found within the 3' non-translated region, coding region and 5' untranslated region. So far two triplet motifs have been observed. CAG and CGG. The numbers give the normal size ranges. In affected individuals the size range of repeats goes from 50 to 2000 in fragile X, 44 to 3000 in myotonic dystrophy and 37 to 121 in Huntington's chorea.

X-LINKED DISORDERS

DNA studies have particular advantages over other methods for detecting female carriers of X-linked disorders because of the problems associated with X-chromosome mosaicism in such females. In each somatic cell of a female one of her two X chromosomes is inactivated. This is often referred to as **Lyonization** after Mary Lyon who first described the phenomenon some twenty years ago. As a result, in a female carrier of an X-linked disorder in some of her cells the active X chromosome will be the one bearing the mutant gene, whereas in other cells the active X chromosome will be the one bearing the normal gene. As X inactivation is a random process in most carriers the distribution of these two types of cells is likely to be about equal. However, there will be occasional carriers in whom most of the

Table 7.3 Gene structure of X-linked disorders

Gene	Disorder	Number of exons
Factor VIII	Haemophilia A	26
Factor IX	Haemophilia B	8
Ornithine carbamoyl transferase	Ornithine carbamoyl transferase deficiency	10
α-Galactosidase	Fabry disease	7
Proteolipid protein	Pelizaeus–Merzbacher	7
Iduronate-2-sulphatase	Mucopolysaccharidosis II (Hunter)	9
btk	Bruton's agammaglobulinaemia	ND
IL2 γR	Severe combined immunodeficiency	8
CD40L	Hyper IgM	5
Cytochrome b245	Chronic granulomatous disease	13
TCD	Choroideraemia	ND
	Menkes	ND
Connexin 32	Charcot–Marie–Tooth	2
Collagen type 4	Alport	51
Dystrophin	Duchenne, Becker muscular dystrophy	79
Emerin	Emery–Dreyfuss muscular dystrophy	ND

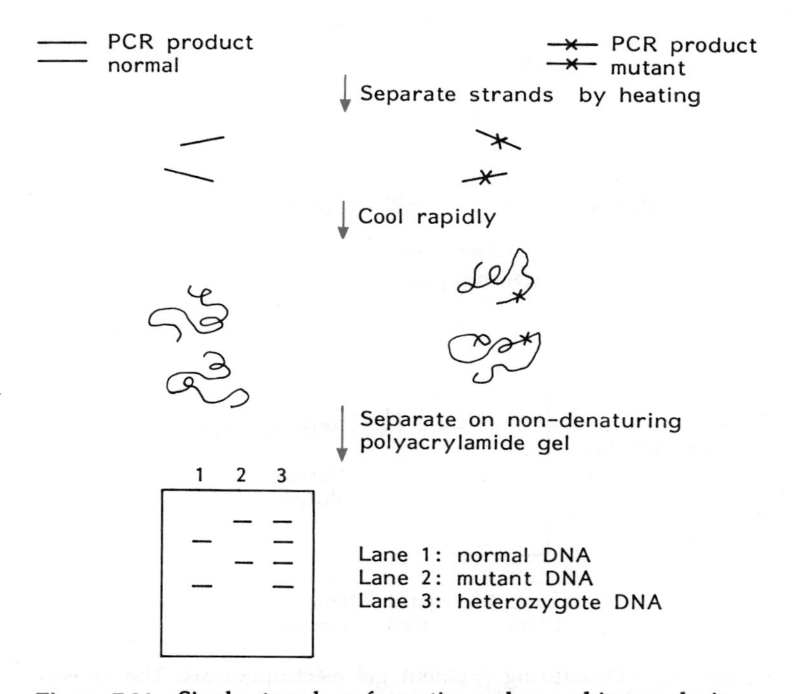

Figure 7.21 Single-strand conformation polymorphism analysis.

active X chromosomes by chance will be those bearing the mutant gene and these women will then exhibit some of the manifestations of the disease: e.g. a bleeding tendency in haemophilia, or enlarged calves and some muscle weakness in Duchenne muscular dystrophy. There will also be occasional carriers in whom most of the active X chromosomes by chance will be those bearing the normal gene. Such carriers could only be detected biochemically by studying single cells or clones of cells.

As explained in the section on Duchenne muscular dystrophy, it is often very important in X-linked disorders to detect the mutation itself and this must be done independently in each family studied. Fortunately the dystrophin gene is not typical of important genes on the X chromosome in its size and complexity (Table 7.3), and sequencing of other X-linked genes is more of a practical proposition, particularly if a preliminary screening procedure is used. One of the most convenient is single-strand conformation polymorphism analysis or SSCP. This works on the principle that single-stranded DNA molecules, obtained by heating a PCR product, will fold into specific structures depending on the sequence and that molecules with different sequences will therefore fold into different three-dimensional structures which will migrate through non-denaturing gels at different

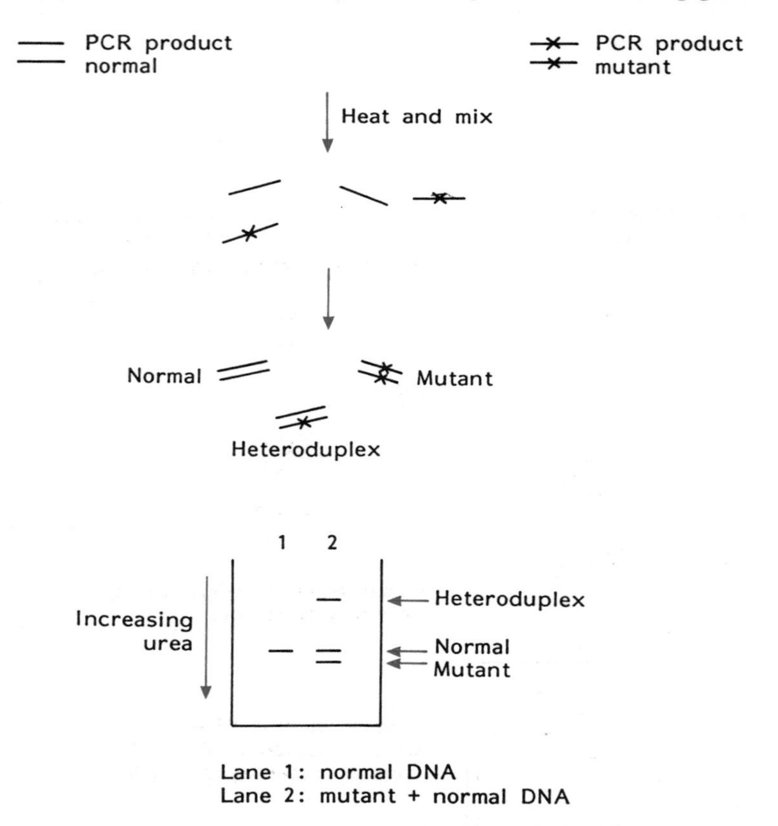

Figure 7.22 Denaturing gradient gel electrophoresis. The heteroduplexes, with a mismatch, will melt and slow down first.

rates (Fig. 7.21). The molecules can be made slightly radioactive by incorporating a small amount of radiolabelled nucleotide into the PCR or a non-radioactive product can be detected after electrophoresis by silver staining. This method is very simple and detects probably around 85% of all point mutations. A slightly more sensitive, but initially more difficult method to set up, is denaturing gradient gel electrophoresis (DGGE). This method makes use of the fact that different DNA sequences will start to melt or breathe at slightly different temperatures and once there is any unfolding of the structure migration through a gel will be drastically reduced. The gel used has a gradient of a denaturing chemical (urea) in it and as a molecule reaches a concentration of urea at which it starts to melt it will slow down, thus distinguishing it from molecules with different sequences (Fig. 7.22). Both these methods have been used effectively to screen quickly for sequence changes in PCR products in order to identify which exon to sequence fully.

EARLIER PRENATAL DIAGNOSIS

Prenatal diagnosis on preimplantation material is seriously inhibited by the very small amount of material available. A way round this has recently been suggested and used in a pilot study. In this method random 15 base pair oligonucleotides (15 mers) are used as primers to amplify, hopefully randomly, the whole genome. The temperature of primer annealing to template is initially very low to encourage random binding. In later rounds the annealing temperature is raised to encourage faithful PCR copying. Subsequently, sequence specific primers are used to look for mutation, e.g. a deletion of the dystrophin gene. Enough initial product should be formed from the random priming to allow several loci to be investigated simultaneously.

CHROMOSOME IDENTIFICATION

Although identification of chromosome abnormalities using banded karyotypes has been available for some time it is often difficult, and requires highly trained operators, to identify the exact origin of a derivative chromosome, particularly in poor quality preparations from leukaemias. This situation has been changed dramatically by the introduction of chromosome 'paints' for use in fluorescent *in situ* hybridization (FISH). The probes for chromosome painting of individual chromosomes are derived from two sources. Either chromosomes are separated using a fluorescence-activated cell sorter (FACS) or the paint is produced from a somatic cell hybrid. FACS sorting of chromosomes depends on the fact that that DNA has affinity for the fluorescent dye ethidium bromide, and the amount of fluorescence produced by an individual chromosome depends on its size. Chromosome preparations are made from peripheral blood lymphocytes by standard methods. The metaphase cells are lysed and the suspended chromosomes stained with ethidium bromide. The stained chromosomes are then subjected to a fine laser beam and sorted by flow cytometry in one of the commercially available instruments. Even though only a very small number of separated chromosomes are produced, chromosome specific sequences can be produced by PCR amplification using primers designed to recognize *Alu* repeat sequences (see page 20). Alternatively, if a somatic cell hybrid containing only one human chromosome is used as the starting material, *Alu* primers will only recognize the human chromosome and not the rodent parent chromosomes. These probes, already labelled, are available commercially. Following the same principles we have described for filter hybridizations, the DNA in metaphase chromosome spreads on microscope slides can be denatured and hybridized. An example is shown in Plate II in which chromosome 4 and chromosome 11 are both painted in a poor quality preparation from a leukaemic cell carrying a 4;11 chromosome translocation. A most elegant adaptation of this involves the whole process in reverse. Metaphase chromosomes are flow sorted to separate the unidentified marker chromosome. This is used as a template to produce *Alu*-PCR products which are in their turn hybridized back to a normal chromosome spread, thus revealing the origin of the two chromosomes contributing to the translocation.

SUMMARY

- A very important practical application of recombinant DNA technology is in the prevention of genetic disease by providing novel and precise methods for detecting preclinical cases in autosomal dominant disorders of late onset and female carriers of X-linked disorders, and for prenatal diagnosis.
- Although most of the approaches currently being used were developed for haemoglobinopathies, they are now applicable to most of the commonest single gene disorders, including cystic fibrosis, muscular dystrophy, Huntington's chorea and the Fragile X form of mental retardation.
- Direct gene mutation is now used in many cases but studies using linked probes are also important.
- The polymerase chain reaction has simplified and speeded up most procedures and even made diagnosis on single cells possible. Fetal chorionic tissue can be obtained during the first trimester of pregnancy and has made early testing of pregnancies at risk a reality.
- Several neurological disorders, particularly those of late onset, have been shown to be caused by expansion of triplet repeats within the genes. This leads to increasing severity in members of the pedigree born in subsequent generations (anticipation).

REFERENCES AND FURTHER READING

Textbooks and review articles

Cao A. *Am J Hum Genet* 1994; 54: 397–402

Bakker E. William Allan Award Address. In: Bartsocas CS, ed. *Genetics of Neuromuscular Disorders*.

Caskey CT, Pizuti A, Fu Y-H, Fenwick RG, Nelson DL. Triplet repeat mutations in human disease *Science* 1992; **256**: 784–789
Collins FS. Cystic fibrosis: molecular biology and therapeutic implications. *Science* 1992; **256**: 774–779
Emery AEH. *Methodology in Medical Genetics*, 2nd edn. Edinburgh and London: Churchill Livingstone, 1986
Emery AEH. *Duchenne Muscular Dystrophy*, 2nd edn. Oxford: OUP, 1993
Galjaard H. *Genetic Metabolic Diseases – Early Diagnosis and Prenatal Analysis*. Amsterdam: Elsevier/North Holland, 1980
Grompe M. The rapid detection of unknown mutations in nucleic acids. *Nature Genetics* 1993; **5**: 111–117
Gusella JF. Elastic DNA – boon or blight? *New Engl J Med* 1993; **329**: 571–572
Sutherland GR, Richards RI. Dynamic mutations on the move. *J Med Genet* 1993; **30**: 978–981
Tsui L-C. The spectrum of cystic fibrosis mutations. *TIG* 1992; **8**: 392–398
Wald NJ, Kennard A. Prenatal biochemical screening for Down's syndrome and neural tube defects. *Current Opinions in Obstetrics and Gynecology* 1992; **4**: 302–307

Research publications

Bakker E, van Broekhoven C, Bonten EJ, van de Vooren MJ, Veenema H, Van Hul W, Van Ommen GJB, Vandenberghe A, Pearson PL. Germline mosaicism and Duchenne muscular dystrophy mutations. *Nature* 1987; **329**: 554–556
Botstein D, White RL, Skolnick M, Davis RW. Construction of a genetic linkage map in man using restriction fragment length polymorphisms. *Am J Hum Genet* 1980; **32**: 314–331
Carter NP, Ferguson-Smith MA, Perryman MT, Telenius H, Pelwear AH, Leversha MA, Glancy MT, Wood SL, Cook K, Dyson HM, Ferguson-Smith ME, Willatt LR. Reverse chromosome painting: a method for the rapid analysis of aberrant chromosomes in clinical cytogenetics. *J Med Genet* 1992; **29**: 299–307
Cavenee WK, Dryja TP, Phillips RA, Benedict WF, Godbout R, *et al*. Expression of recessive alleles by chromosomal mechanisms in retinoblastoma. *Nature* 1983; **305**: 779–784
Chamberlain JS, Gibbs RA, Ranier JE, Nguyen PN, Caskey C. Deletion screening of the Duchenne muscular dystrophy locus via multiplex DNSA amplification. *Nucleic Acids Res* 1988; **16**: 11141–11156
Chelly J, Concordet J-P, Kaplan J-C, Kahn A. Illegitimate transcription: transcription of any gene in any cell type. *Proc Natl Acad Sci USA* 1989; **86**: 2617–2621
Darras BT, Francke U. A partial deletion of the muscular dystrophy gene transmitted twice by an unaffected male. *Nature* **329**: 556–558
Conner BJ, Reyes AA, Morin C, Itakura K, Teplitz RL, Wallace RB. Detection of sickle cell β-globin allele by hybridization with synthetic oligonucleotides. *Proc Natl Acad Sci USA* 1983; **80**: 278–282
Cuckle HS, Wald NJ, Thompson SG. Estimating a woman's risk of having a pregnancy associated with Down's syndrome using her age and serum alpha-fetoprotein level. *Br J Obstet Gynaecol* 1987; **94**: 387–402
Harper PS. *Practical Genetic Counselling*, 3rd edn. Bristol: John Wright, 1988
Kerem B, Rommens JM, Buchanan JA, Markiewicz D, Cox TK, Chakravarti A, Buchwald A, Tsui LC. Identification of the cystic fibrosis gene: genetic analysis. *Science* 1989; **245**: 1073–1080
Koenig M, Hoffman EP, Bertelson CJ, Monaco AP, Feener C, Kunkel LM. Complete cloning of the Duchenne muscular dystrophy (DMD) cDNA and preliminary genomic organization of the DMD gene in normal and affected individuals. *Cell* 1987; **50**: 509–517
Kristjansson K, Chong SS, Van en Veyver IB, Subramanian S, Snabes MC, Hughes MR. Preimplantation single cell analyses of dystrophin gene deletions using whole genome amplification. *Nature Genetics* 1994; **6**: 19–23
Kunkel LM, Monaco AP, Middlesworth W, Ochs HD, Latt SA. Specific cloning of DNA fragments absent from the DNA of a male patient with an X chromosome deletion. *Proc Natl Acad Sci USA* 1985; **82**: 4778–4782

Lathrop GM, Lalouel JM. Easy calculations of lod scores and genetic risks on small computers. *Am J Hum Genet* 1984; **36**: 460–465

Monaco AP, Bertelson CJ, Liechti-Gallati S, Moser H, Kunkel LM. An explanation for the phenotypic differences between patients bearing partial deletions of the DMD locus. *Genomics* 1988; **2**: 90–95

Old JM, Varawella NY, Weatherall DJ. Rapid detection and prenatal diagnosis of β-thalassaemia: studies in Indian and Cypriot populations in the UK. *Lancet* 1990; **336**: 834–837

Orkin SH, Kazazian HH, Antonarakis SE, Goff SC, Boehm CD, *et al*. Linkage of β-thalassaemia mutations and β-globin gene polymorphisms with DNA polymorphisms in human β-globin gene cluster. *Nature* 1982; **296**: 627–631

Orr HT, Chung M, Banfi S, Wiatowski TJ, Servadio A, Beaudet AL, McCall AE, Duvick LA, Ranum LPW, Zoghbi HY. Expansion of an unstable trinucleotide repeat CAG repeat in spinocerebellar ataxia type I. *Nature Genetics* 1993; **4**: 221–225

Ott J. Estimation of the recombination fraction in human pedigrees; efficient computation of the likelihood for human linkage studies. *Am J Hum Genet* 1974; **26**: 588–597

Ray PN, Belfall B, Duff C, *et al*. Cloning the breakpoint of an X;21 translocation associated with Duchenne muscular dystrophy. *Nature* 1985; **318**: 672–675

Riordan JR, Rommens JM, Kerem B, Alon N, Rozmahel R, Zbyszki G, Zielenski J, Lok S, Plavsic N, Chou J, Drumm M, Ianuzi M, Collins FS, Tsui L-C. Identification of the cystic fibrosis gene: cloning and characterisation of complementary DNA. *Science* 1989; **245**: 1066–1073

Rodeck CH, Morsman JM. First-trimester chorion biopsy. *Br Med Bull* 1983; **29**: 338–342

Rommens JM, Ianuzzi MC, Kerem B, Drumm ML, Melmer G, Dean M, Rozmahel R, Cole J, Kennedy D, Hidaka N, Zsiga M, Buchwald M, Riordan JR, Tsui J-C, Collins FS. Identification of the cystic fibrosis gene: chromosome walking and jumping. *Science* 1989; **245**: 1059–1065

Saiki RK, Walsh PS, Levenson CH, Erlich HA. Genetic analysis of amplified DNA with immobilized sequence-specific oligonucleotide probes. *Proc Natl Acad Sci USA* 1989; **86**: 6230–6234

Wald NJ, Cuckle HS. Biochemical screening. In: Brock DJH, Rodeck CH, Ferguson-Smith MA, eds. *Prenatal Diagnosis and Screening*. Churchill Livingstone, 1992, pp. 563–577

Woo SLC, Lidsky AS, Guttler F, Chandra T, Robson KJH. Cloned human phenylalanine hydroxylase gene allows prenatal diagnosis and carrier detection of classical phenylketonuria. *Nature* 1983; **306**: 151–155

Yau SC, Roberts RG, Bobrow M, Mathew CG. Direct diagnosis of carriers of point mutations in Duchenne muscular dystrophy. *Lancet* 1993; **341**: 273–275

Zhang L, Cui X, Schmitt K, Huber R, Navidi W. Whole genome amplification from a single cell: implications for genetic analysis. *Proc Natl Acad Sci USA* 1992; **89**: 5847–5851

Chapter 8
Treatment

The new technology has so far found most application in the prevention of genetic disease. But it also offers new approaches to treatment which no doubt will be increasingly exploited in the years to come. These can be conveniently discussed under three main headings: biosynthesis, gene therapy and improved diagnosis. The former involves inserting a human gene which codes for a therapeutically important peptide into a suitable vector which is then cloned. In this way the peptide can be synthesized in significant quantities and should prove cheaper and safer than the usual laborious alternative of extraction and purification from human tissues. Several peptides of therapeutic importance have already been successfully synthesized in this way and are summarized in Table 8.1.

Gene therapy, on the other hand, is concerned with treating genetic disease by the replacement or correction of a defective mutant gene. Although enormous advances have occurred in the scientific basis for gene therapy, clinical applications have been slower to follow because of the enormous complexities involved. However, treatment by gene therapy is regarded as something of a holy grail and the several clinical trials now authorized will hopefully soon yield good results.

BIOSYNTHESIS

It has already been seen that there are several important differences in gene structure and function between prokaryotes and eukaryotes (page 63). Because of these differences, certain requirements are necessary for the successful expression of a human gene in, say, *E. coli*. Firstly, the human DNA sequence to be inserted should not contain introns because prokaryotes do not have the necessary enzymes for RNA splicing. This can be achieved by inserting either a chemically synthesized DNA sequence without introns or a cDNA sequence derived from mRNA. Secondly, it is necessary to insert a prokaryote promoter upstream from the gene to be expressed since these are somewhat different in eukaryotes. There are several such promoters – *lac* (lactose metabolism) and *trp* (tryptophan metabolism) genes in *E. coli*, the β-lactamase (ampicillin resistance) gene of the plasmid pBR322 and the erythromycin resistance gene of *Bacillus subtilis*. Of these the *lac* promoter has so far been most widely used. This is part of the so-called *lac* operon which, put very simply, consists of a promoter, an operator and adjacent β-galactosidase and related genes which the operator gene controls. The operator is activated (derepressed) when an inducer (the metabolite lactose in the case of the *lac* operon)

Table 8.1 Products produced by recombinant DNA on the market

Product	*Launch date*
Insulin	1982
Growth hormone	1985
Interferons (various)	1986
Tissue plasminogen activator	1987
Erythropoietin	1989
Granulocyte stimulating factor	1991
Granulocyte macrophage colony stimulating factor	1991
Hepatitis B vaccine (evaluated clinically but not widely available)	1986
Products in phase 2 development	
DNase I	
CD4	
Tumour necrosis factor	
Epidermal growth factor	
Atrial natruretic peptide	

is present. Thus when lactose is present the operator is activated which then turns on the β-galactosidase gene which breaks down the lactose into glucose and galactose. Thus by manipulating the substrate concentration of lactose the activity of the operon can be controlled. Vectors containing various parts of the *lac* operon have been constructed over the years, a good example being λ ρ *lac* derived from a phage and containing the *lac* promoter, operator and most of the β-galactosidase gene.

A few examples of the various strategies employed to synthesize medically important peptides using recombinant DNA technology will illustrate some general points.

Somatostatin

Somatostatin is a peptide hormone which, among other things, inhibits growth hormone and is used in the treatment of children with excessive growth. It was the first medically important peptide to be successfully synthesized using recombinant DNA technology. The peptide consists of 14 amino acids, and from knowing the amino acid structure it is possible to infer the nucleotide base composition of its gene.

Itakura and colleagues in 1977 put together the 14 codons for the structural information in the somatostatin gene (using where possible *E. coli*-preferred codons) to which they also added several other features. Included was a methionine (ATG) start codon preceding the first amino acid codon, and two nonsense (TGA, TAG) stop codons following the last amino acid codon. They also included single-stranded cohesive termini for the enzyme *Eco*RI at one end and for the enzyme *Bam*HI at the other end so as to facilitate its insertion at these sites in the plasmid vector pBR322. Upstream from this constructed somatostatin DNA fragment were inserted elements of the *lac* operon.

This construction leads to the synthesis of a hybrid protein consisting of part of the β-galactosidase protein fused to the peptide hormone. The active hormone is cleaved from the hybrid protein using cyanogen bromide (CNBR) which specifically cleaves peptides at methionine, which in this case lies between the two molecules. The synthesis of peptides whose release depends on CNBR cleavage is clearly limited to those which do not contain methionine.

The various steps involved in generating a recombinant plasmid for the bacterial synthesis of somatostatin are summarized in Fig. 8.1.

Similar strategies have also been used for the synthesis of other small peptides such as the opioid peptides (endorphins and enkephalins), which act mainly as inhibitors in the sensory, neuroendocrine and autonomic systems, and which are therefore likely to have therapeutic value in painful and stressful conditions.

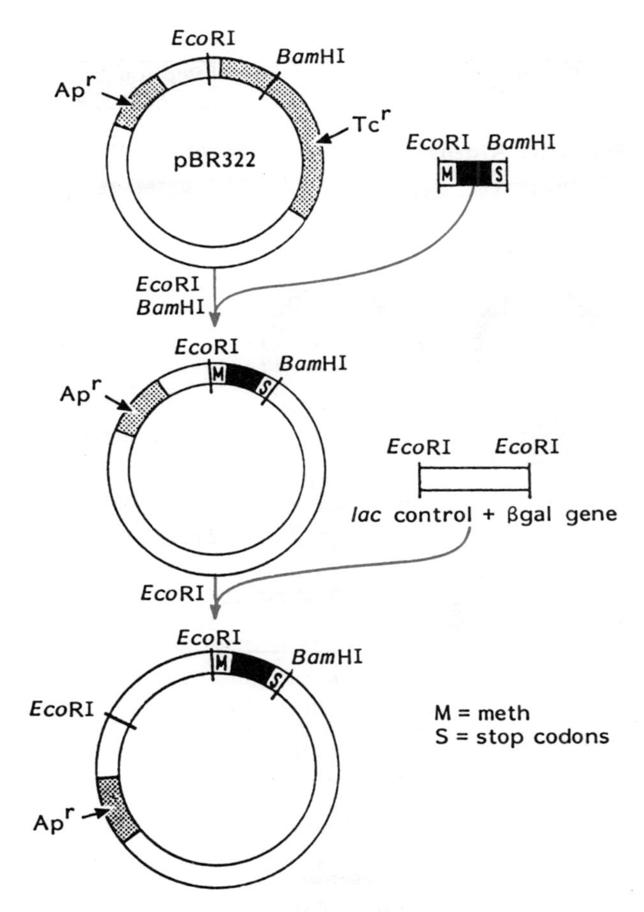

Figure 8.1 Simplified diagrammatic representation of the steps involved in generating a recombinant plasmid for the bacterial synthesis of somatostatin.

Insulin

Insulin is essential for the treatment of the more severe (insulin-dependent) forms of diabetes mellitus. In the past this was usually obtained from beef or pig pancreas but since these insulins are not chemically identical to human insulin, a proportion of diabetics may produce antibodies which can seriously interfere with treatment. The biosynthesis of human insulin is therefore of considerable importance.

The molecule consists of two peptide chains held together by two disulphide bonds: an A chain of 21 amino acids and a B chain of 30 amino acids. The genes for the two peptides are synthesized from the inferred codons, each with *Bam*HI and *Eco*RI cohesive ends. These are then inserted into the plasmid pBR322 along with elements of the *lac* operon and cloned **separately** (Fig. 8.2). As in the case of somatostatin, the hybrid proteins are each cleaved from β-galactosidase with CNBR. The two peptides are then joined by disulphide bonds by sulphonation and air oxidation under appropriate conditions.

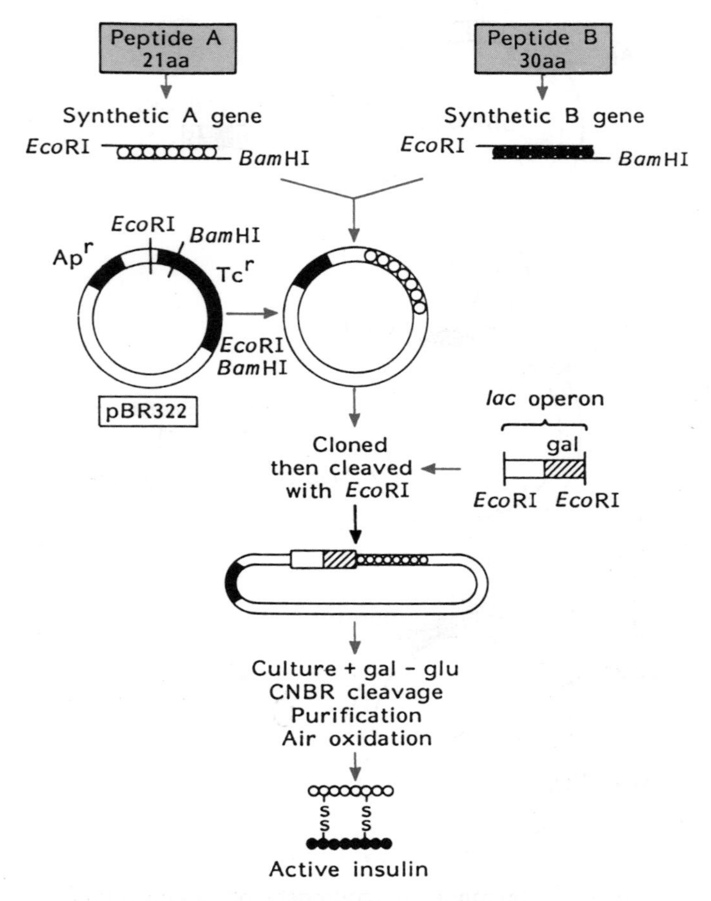

Figure 8.2 Simplified diagrammatic representation of the bacterial synthesis of human insulin.

Human insulin produced in this way by recombinant DNA technology is quite as effective as porcine insulin and is without any apparent side-effects. It is now being produced commercially and marketed under the name Humulin by the Eli Lilly Pharmaceutical Company.

Human growth hormone

The peptides discussed so far contain relatively few amino acids, and therefore the chemical synthesis of the genes coding for them is feasible. However, with large peptides this becomes very difficult and time-consuming and so resort has to be made to cloning cDNA prepared from mRNA extracted from a relevant tissue.

Human growth hormone (HGH), used in the treatment of children with growth retardation, occupies an intermediate position in this regard. The reason is that though the peptide contains 191 amino acids, it can be synthesized in two fragments which are then joined together. The latter is made possible because HGH cDNA has an *Hae*III site in the sequence coding for amino acids 23 and 24, and the cohesive ends at this point can be used to join the two fragments. Essentially the method involves extracting mRNA for HGH from pituitary tissue from which cDNA was made. This was then cleaved with *Hae*III. The smaller fragment was discarded and the larger fragment (codons for amino acids 24 to 191) was retained and cloned in a plasmid. The smaller fragment was synthesized from its constituent nucleotides and cloned. The two fragments were then digested at their *Hae*III sites and the fused gene, with an *Eco*RI cohesive site at one end and an *Sma*I blunt-ended site at the other was inserted downstream from two *lac* promoters in tandem in the plasmid pGH6 and cloned (Fig. 8.3).

As successful as this method has been, the product is not completely identical to the hormone as secreted by the pituitary. It has an extra NH_2-terminal methionine and for this reason could be antigenic, though it is as potent as the pituitary hormone itself. One answer to the problem has been to generate an SV (Simian virus) 40 recombinant virus containing the entire cDNA sequence controlled by the SV40 promoter (a strong promoter) using monkey kidney cell culture as the host. These cells excrete HGH identical to that isolated from pituitary glands.

The successful biosynthesis of HGH is extremely important for several reasons. In the past the only source of HGH was human pituitary tissue removed at autopsy. A child with pituitary dwarfism usually needs twice-weekly injections until the age of 20, a regime which would require over a thousand pituitaries. In the past autopsy sources could only just keep up with the demand and tragically, many children treated have subsequently developed Creutzfeldt–Jacob disease because of small amounts of contaminating virus. The ready availability of bioengineered HGH means that other possible therapeutic uses can now be explored, such as its use in speeding up healing processes in bone fractures (especially in the elderly), after extensive burns and following major surgery.

Eukaryotic cloning systems

Most of the emphasis so far has been on prokaryotic cloning systems employing *E. coli* as the host organism. Some of the problems of using such a system for the

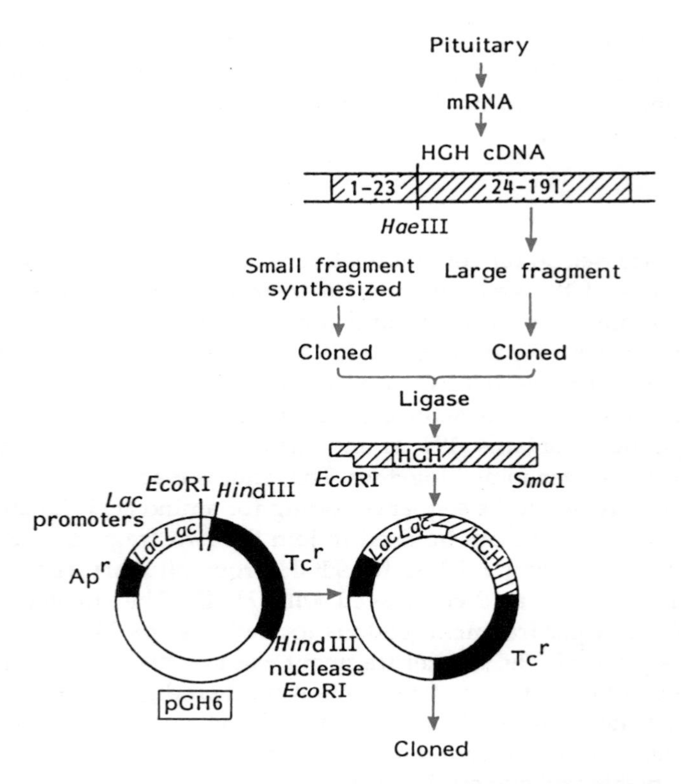

Figure 8.3 Simplified diagrammatic representation of the steps involved in generating a recombinant plasmid for the bacterial synthesis of human growth hormone (HGH).

expression of eukaryotic DNA have already been mentioned, along with possible solutions. However, even when a peptide has been synthesized correctly there may still be problems. The peptide may prove toxic to the host prokaryote or may even be destroyed by it. Further, in eukaryotes proteins are subjected to a number of post-translational modifications which may affect their activity or stability, modifications which do not occur in prokaryotes. For these, and other reasons, various eukaryotic host–vector systems have been developed. Vectors developed include SV40, adenovirus, adeno-SV40 hybrid viruses, bovine papillomavirus and retroviruses. Eukaryotic hosts include yeast and various mammalian (e.g. mouse, monkey, human) cultured cells. It would not be appropriate to go into these matters in detail, but one example will illustrate the possible potential. Plasmids have been constructed which direct the synthesis of human interferon in yeast. The construction is a very clever one and is derived from the plasmid pBR322 (Fig. 8.4).

It contains an ampicillin resistance gene derived from pBR322, a yeast promoter (from the gene for 3-phosphoglycerate kinase or PGK) and TRP gene which produces tryptophan and thus permits selection for the plasmid in yeast cells growing in a medium lacking tryptophan. Interferon genes can be inserted as *Eco*RI

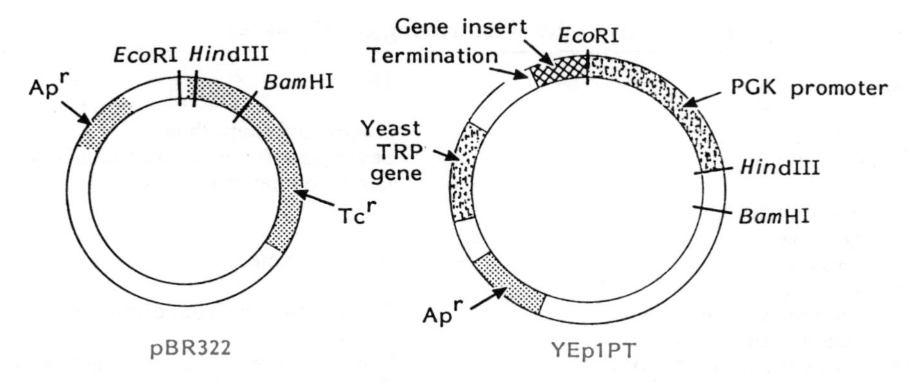

Figure 8.4 Plasmids pBR322 and YEp 1PT used respectively with *E. coli* and yeast as the host organisms.

restriction fragments into the only *Eco*RI site. A somewhat similar plasmid has also been employed for synthesizing human hepatitis B antigen in yeast.

Transgenic animals

A promising method for production of recombinant human protein is in the milk of transgenic livestock. A suitable human gene construct is injected into a fertilized egg which is then reimplanted into a female. Hopefully, some of the offspring born will have inserted the gene into their chromosomes. If the insertion has occurred in germ cells then future generations can be bred carrying the inserted gene stably in every cell. The protein coding sequences are attached to control sequences and a promoter for the milk protein gene β-lactoglobulin, which means that the introduced gene should be expressed in mammary glands and the protein product recoverable from milk. Sheep and cows will both secrete large amounts of the desired protein, up to 5 g per litre, in their milk and milk production is, of course, a well established farming practice in these livestock. α_1-Antitrypsin, factor VIII and glucocerebrosidase (see below) have all been produced in limited quantities in sheep.

GENE THERAPY

Some genetic disorders can already be treated through a variety of strategies (Table 8.2), and in Gaucher's disease attempts have even been made to replace the deficient enzymes.

So far this has been directed towards altering the phenotype – so-called 'phenotypic engineering'. In many cases such treatment has only been partially successful, sometimes prohibitively expensive and often very difficult to institute and maintain. The new technology offers an entirely new approach which might one day replace other methods. The history of the development of gene therapy has been outlined by Friedmann (1992).

Table 8.2 Examples of various methods for treating genetic disease

Therapy	Disorder
1. Gene therapy	Haemoglobinopathies
Cloned gene (P)	Severe-combined immunodeficiency
	Lesch–Nyhan syndrome
2. Enzyme induction	
(a) By virus	
Shope virus (P)	Argininaemia
(b) By drugs	
Phenobarbitone	Congenital non-haemolytic jaundice
3. Replacement of deficient enzyme	
(a) Tissue transplantation (S)	Fabry's disease, mucopolysaccharidoses
(b) Enzyme preps: trypsin	Trypsinogen deficiency
alpha-1-antitrypsin	Alpha-1-antitrypsin deficiency
4. Replacement of deficient protein	
Antihaemophilic globulin	Haemophilia
5. Replacement of deficient vitamin or coenzyme	
B6	Cystathioninuria
B12	Methylmalonicacidaemia (S)
Biotin	Propionicacidaemia (S)
D	Vitamin D-resistant rickets
6. Replacement of deficient product	
Cortisone	Adrenogenital syndrome
Cysteine	Homocystinuria
Thyroxine	Congenital cretinism
Uridine	Oroticaciduria
7. Substrate restriction in diet	
(a) Amino acids	
Phenylalanine	Phenylketonuria
Leucine, isoleucine and valine	Maple syrup urine disease
Methionine	Homocystinuria
(b) Carbohydrate	
Galactose	Galactosaemia
(c) Lipid	
Cholesterol	Hypercholesterolaemia
(d) Total protein	Defects in urea cycle
8. Induction of alternative pathways	
Citric acid (P)	Defects in urea cycle
9. Drug therapy	
Aminocaproic acid	Angioneurotic oedema
Cholestyramine	Hypercholesterolaemia
Insulin	Diabetes
Pancreatin	Cystic fibrosis
Penicillamine	Wilson's disease
Probenecid, allopurinol	Hyperuricaemia
Antibiotics	Various metabolic disorders
10. Preventive therapy	
Avoidance of certain drugs	G6PD deficiency, porphyria
Rh gamma globulin	Rh incompatibility
11. Replacement of defective tissue	
Kidney transplantation	Polycystic kidney disease
Corneal graft	Congenital keratoconus

Table 8.2 *(continued)*

Therapy	Disorder
12. Removal of diseased tissues	
Colectomy	Polyposis coli
Splenectomy	Hereditary spherocytosis
Neurofibromas	Neurofibromatosis
13. Portacaval anastomosis (P)	Hepatic glycogenoses
	Hypercholesterolaemia
14. Extracorporeal removal of lipids (P)	Hypercholesterolaemia

P = possible but not yet proven.
S = effective in some cases.

Gaucher's disease is a recessive lysosomal storage disorder caused by a deficiency of the lysosomal hydrolase glucocerebrosidase. In the absence of the enzyme the breakdown of the substrates is prevented and they accumulate. The accumulation of the trapped intermediates results in a storage disease. Glucosylceramide accumulates in monocyte/macrophage cells causing disease manifestations in the spleen, liver, bone marrow and other visceral organs. The severity varies widely, depending on the gene mutation and patients who are homozygous for a N370S mutation (mainly Ashkenazi Jews) have the mildest type I form of disease. The risks of bone marrow transplantation do not justify the procedure in patients with mild disease and replacement of the deficient enzyme with glucocerebrosidase extracted from human placental extracts has been used. Receptor mediated uptake into macrophage cells depends on the presence of the correct carbohydrate side chains, in this case a mannose residue. Native glucocerebrosidase is poorly taken up but modification of carbohydrate residues by sequential enzymatic deglycosylation greatly increases the uptake and led to the industrial preparation of the drug alglucerase or Ceredase. The replacement enzyme alleviates the symptoms of the milder, non-neuropathic form, and is well tolerated. However, the financial cost is enormous, the annual cost for the enzyme being in the order of $400 000. Although recombinant enzyme produced in Chinese hamster ovary cells is now being produced the cost is unlikely to be much less and this disorder, along with many others, will ultimately depend on gene therapy for treatment of most patients.

It would be expected that when there is a gene deletion therefore no gene product, treatment with the missing enzyme or factor would result in the production of antibodies to the therapeutic agent. This would arise because the individual's immune system would never have experienced the gene product and would therefore view such substances as 'non-self'. In fact it has been known for some time that a proportion of patients with inherited growth hormone deficiency and antihaemophilic factor IX deficiency do produce antibodies when treated, and this poses a very serious problem in management. At least some of these individuals have now been shown to have deletions of the responsible gene. To generalize, the development of antibodies which inhibit a therapeutic agent in any inherited metabolic disorder would suggest that the underlying molecular basis for the disorder may well be a gene deletion. Conversely, the detection of a gene deletion in an individual affected with a metabolic disorder could significantly affect the approach to treatment.

DEVELOPMENTS IN GENE THERAPY

The earliest experiments to test the first hypotheses of gene therapy, i.e. that cells could be stably transformed with foreign DNA and a missing genetic function restored, used mutagenized cell lines transfected with total DNA introduced by calcium phosphate transfection. Progress has been extremely rapid since then and a variety of methods for introducing foreign DNA into cells are now favoured, the method of choice usually depending on the most appropriate tissue for delivery. Some examples in current development are given in Table 8.3.

Retroviruses

The apparent advantage of retroviral vectors is that they can stably integrate into the DNA of the target cells following *ex vivo* application (see Fig. 6.7). This involves removal of the relevant target cells from the body, transduction of the cells *in vitro*, and subsequent reintroduction of the modified cells into the patient. However, the efficiency of transduction is dependent on the recipient cells having the correct viral receptors and on their ability to divide in order for the viral vector to be integrated. Progress in retroviral mediated gene transfer has depended on the development of replication defective cell lines (packaging cells) which lack the information for virus production and, theoretically pose no safety problems. Diseases affecting hematopoietic cells have been an obvious choice because of the possibility of transducing stem cells in the bone marrow and thus providing a continual supply of corrected cells. Unfortunately the stem cells within bone marrow are not well defined yet. For once haemoglobinopathies have not led the way. This is because

Table 8.3 Some examples of current developments in gene therapy

Method of delivery	Target	Examples	Disorder
Retrovirus	Bone marrow	Adenosine deaminase	Severe combined immunodeficiency
	Hepatocytes	Low density lipoprotein receptor	Hypercholesterolaemia
	Skeletal muscle	Dystrophin mini-gene	Mdx mouse
Adenovirus	Nasal epithelial cells	CFTR	Cystic fibrosis
	Lung epithelial cells	α_1-antitrypsin	α_1-antitrypsin deficiency
	Skeletal muscle cells	Dystrophin mini-gene	Duchenne muscular dystrophy
Liposomes	Nasal epithelial cells	CFTR	Cystic fibrosis
Herpes simplex virus	Central nervous system		

the ratio of α-globin to the β-globin chain is crucial and this is not readily controlled. However, the human adenosine deaminase gene has been replaced in a number of children with severe combined immunodeficiency. The vector used is shown in Fig. 8.5. A further problem with retroviral vectors is that there is a limit on the size of the insert that can still be successfully packaged. In the case of dystrophin, as the full length cDNA is 14 kb the insert would be too long for the virus to package. A most ingenious solution to this problem has been devised. A patient with mild Becker muscular dystrophy was found with a very large in-frame deletion (see page 144) consisting of over 40% of the central rod domain of dystrophin. This mini cDNA of only 6.3 kb is small enough to be packaged and has been used as the basis for a construct which has successfully transduced skeletal muscle tissues of the mdx mouse (a mouse model of Duchenne muscular dystrophy with a mutation of the mouse dystrophin gene).

Hepatocytes have been shown to be susceptible to retroviral infection with potential uses in the treatment of major disorders such as PKU, blood clotting and α_1-antitrypsin deficiency. An apparently major problem is how to transplant sufficient hepatocytes back into the patient. However, it was demonstrated in animal experiments that hepatocytes could be reinjected via the portal venous system, the veins which drain from the intestine into the liver, and can seed in the liver from there. This method has been used to transfer the low-density lipoprotein receptor (LDLR) gene into a patient who was homozygous for familial hypercholesterolaemia due to a missense mutation of the LDLR gene, with initially promising results.

Adenovirus vectors

The advantage of using adenoviruses as vectors is their ability to infect a wide range of cells without division being necessary. They are particularly suitable for airway epithelial cells, as they are a common cause of upper respiratory tract infection in humans, and have therefore been a focus of attention for cystic fibrosis gene therapy. Clinical trials are taking place in the USA although one early trial was halted when a patient developed an inflammatory response. The obvious

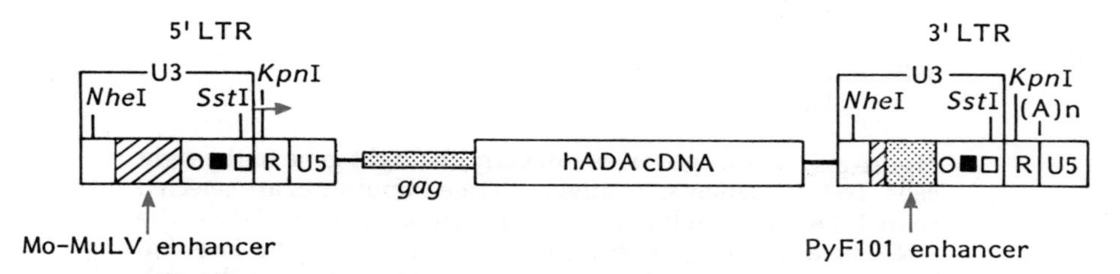

Figure 8.5 Replication defective retrovirus: pLgAL (ΔMo + PyF101).
- Standard MoMuLV based construct.
- Expression of hADA driven by hybrid LTR (ΔMo + PyF101).
- Single cDNA insert, no selectable marker.
- Previously used for long-term reconstitution studies in rhesus monkeys.

disadvantage is that the virus does not stably integrate and regular reapplication will be necessary. On the other hand, this could be seen as an advantage in some ways as there have been doubts expressed about the safety of viral integration into the genome at random sites which could in the worst case activate an oncogene. Adenovirus vectors seem more promising than retroviral ones in the treatment of DMD.

Liposomes

A cystic fibrosis trial is also being carried out using delivery via liposomes. These are certainly safe as they have been used for many years by the pharmaceutical industry and as they can be injected through the cell membrane they are a suitable method of delivery to skin, nasal or muscle cells (Fig. 8.6).

Myoblast transfer

A very novel approach to the therapy of muscular dystrophy has been introduced by Partridge and colleagues in which they injected normal myoblasts into the muscle of the dystrophic mdx mouse described above. Many of the fibres around the site of injection were rendered dystrophin positive presumably because the normal myoblasts fused with host muscle fibres. But so far trials in humans have

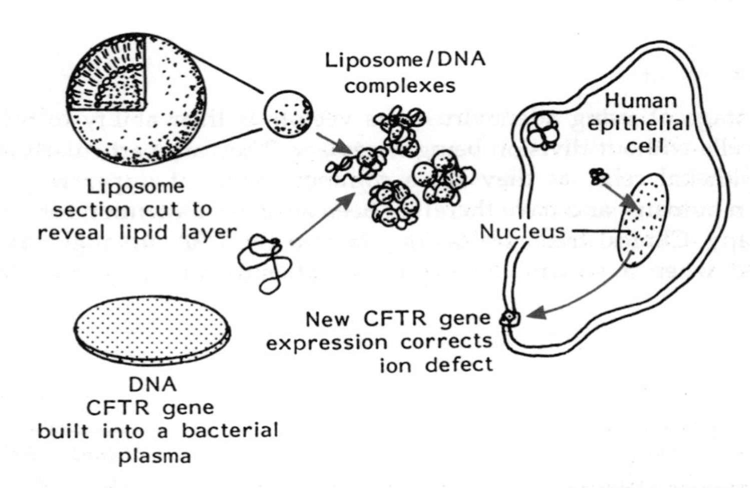

Figure 8.6 How liposomes can carry the right gene to the right cells. To treat patients, positively charged liposomes are selected so that they will combine with the negatively charged DNA. The DNA of a bacterial plasmid has had the CFTR gene inserted. The complex fuses with the cell membrane, entering the cell, where the DNA is released and transported to the nucleus. Here the new gene is inserted into the cell's DNA, so that the normal protein is produced, correcting the ion transport defect which causes cystic fibrosis. (Courtesy Eddie Kinrade, from *MRC News* with permission.)

not been encouraging and this seems unlikely to prove an effective approach to treatment.

THERAPIES BASED ON RNA

Therapies based on antisense oligonucleotides or RNA are being investigated enthusiastically although practical applications will probably not be common for some time. They are based on the idea that a suitably designed antisense molecule can inhibit a specific sequence to which it is complementary and can inhibit translation of a given messenger RNA either by blocking ribosomal binding or by activating RNase H (a ribonuclease which degrades double-stranded RNA specifically). An example will illustrate the principle. As already explained, in chronic myelogenous leukaemia there is a translocation between the chromosomes 9 and 22 resulting in the production of a hybrid mRNA (and subsequently protein) BCR-ABL (see Fig. 6.9). Calabretta and his group isolated the BCR-Abl mRNA from a particular tumour by RT-PCR (see page 51) and determined the exact DNA sequence across the junction between the BCR sequence (from 22) and ABL (from 9). They synthesized an 18 bp oligonucleotide complementary to this junction and exposed leukaemia blast cells to the synthetic oligonucleotide. The hybrid oligonucleotide will not bind to the normal ABL or BCR genes but only to the fused mRNA. In colony assays blast cells were specifically inhibited showing that one day this may form the basis for gene-targeted selective killing of neoplastic cells.

Another approach utilizes the catalytic activity of certain RNAs by designing them to bind specifically to certain mRNAs and then cleave, and thus inactivate, them. These are called ribozymes and their use has been proposed to destroy viral RNAs, activated oncogene mRNAs or HIV mRNA.

RESIDUAL DISEASE

A last example of the application of recombinant DNA to treatment is really the reverse, in fact the extremely sensitive testing for residual disease in a cancer patient in remission, so that if residual disease is absent the patient can be saved further treatment. Patients with acute promyelocytic leukaemia (PML) have a fusion transcript of the PML gene (from chromosome 15) and the retinoic acid receptor (RAR) (from chromosome 17). It is possible to design primers for the PCR, one on each side of the breakpoint, which will identify this fused mRNA transcript by RT-PCR. They will, of course, pick up no transcript from normal chromosomes as the corresponding ends will be on separate chromosomes and not attached. APL is responsive to chemotherapy but some patients who appear to be in complete remission eventually relapse. Patients were tested between 2 and 21 months after complete remission: of 13 who were PML-RAR positive 11 relapsed within months but of the 12 who were negative none went on to develop disease. Therefore, a negative RT-PCR result indicates that it is not necessary for the patient to undergo further treatment.

PRESYMPTOMATIC TESTING

Finally, an example along the same lines, is the presymptomatic testing in families where individuals are at risk of developing tumours, e.g. multiple endocrine neoplasia or colonic cancer. If the individuals in a family can be divided into those at risk and those not at risk, suitable regular screening, with subsequent treatment by surgery, can be offered to those at risk whereas those not at risk can be reassured and avoid unpleasant and costly screening procedures.

SUMMARY

- Recombinant DNA technology is providing new approaches to treatment through the biosynthesis of therapeutically important products and eventually through gene therapy.
- The former has been achieved by inserting a gene which codes for a particular peptide, along with an appropriate promoter, into a suitable vector which is then cloned. In this way it has been possible to synthesize, for example, somatostatin, insulin, growth hormone, interferon, opioid peptides (enkephalins and endorphins), interleukins, thymosin, urokinase and haemophilic factor VIII.
- Gene therapy is increasingly seen as a realistic possibility. Genes could be stably introduced into cells using retroviruses but this depends on knowledge of suitable stem cells. There are worries over the safety of inserting retrovirus sequences at random into the genome.
- Other approaches include adenovirus vectors or liposome delivery of genes. Either of these approaches would only result in transient expression.
- For ethical and various other reasons, embryo treatment seems very unlikely in humans.

REFERENCES AND FURTHER READING

Textbooks and review articles

Anderson WF. Human gene therapy. *Science* 1992; **256**: 808–813

Barinaga M. Ribozymes: killing the messengers. *Science* 1993; **262**: 1512–1514

Beutler E. Gaucher disease: new molecular approaches to diagnosis and treatment. *Science* 1992; **256**: 794–798

Clarke AJ, Simmons JP, Wilmut I, Lathe R. Pharmaceuticals from transgenic livestock. *Trends Biotech* 1987; **5**: 20–22

Emery AEH. *Methodology in Medical Genetics*, 2nd edn. (1986). Edinburgh and London: Churchill Livingstone

Friedmann T. A brief history of gene therapy. *Nature Genetics* 1992; **2**: 93–98

Harris TJR. Expression of eukaryotic genes in *E. coli*. In: Williamson R, ed. *Genetic Engineering*, Vol. 4, London: Academic Press, 1983, pp. 127–185

Miller D. Human gene therapy comes of age. *Nature* 1992; **357**: 455–460
Mulligan RC. The basic science of gene therapy. *Science* 1993; **260**: 929–931

Research publications

Barton NW, Brady RO, Dambrosia JM, Di Bisceglie AM, Doppelt SH, Hill SC, Mankin HJ, Murray GJ, Parker RI, Argoff CE, Grewal RP, Yu K-T and colleagues. Replacement therapy for inherited enzyme deficiency-macrophage targeted glucocerebrosidase for Gaucher's disease. *N Engl J Med* 1991; **324**: 1464–1470

Beutler E. Gaucher disease: New molecular approaches to diagnosis and treatment. *Science* 1992; **256**: 794–799

Grossman M, Roper SE, Kozarzky K, Stein EA, Engelhardt JF, Muller D, Lupien PJ, Wilson JM. Successful ex vivo gene therapy directed to liver in a patient with familial hypercholesterolemia. *Nature Genetics* 1994; **6**: 335–341

Itakura K, Hirose T, Crea R, Riggs AD, Heyneker HL, *et al.* Expression in *E coli* of a chemically synthesized gene for the hormone somatostatin. *Science* 1977; **198**: 1056–1063

Jackson DA, Symons RH, Berg P. Biochemical method for inserting new genetic material into DNA of simian virus 40: Circular SV40 DNA molecules containing lambda phage genes and the galactose operon of *Escherichia coli*. *Proc Natl Acad Sci USA* 1972; **69**: 2904–2909

Kinrade E. *MRC News* 1994; **62**: 19

Ledley FD, Darlington GJ, Hahn T, Woo SCL. Retroviral gene transfer into primary hepatocytes: Implications for genetic therapy of liver specific functions. *Proc Natl Acad Sci USA* 1987; **84**: 5335–5339

Lo Coco F, Diverio D, Pandalfi PP, Biondi A, Rossi V, Avvisati G, Rambaldi A, Arcese W, Petti MC, Meloni G, Mandelli F, Grignani F, Masera G, Barbui T, Pelicci PL. Molecular evaluation of residual mRNA as a predictor of relapse in promyelocytic leukaemia. *Lancet* 1992; **340**: 1437–1439

Mann R, Mulligan RC, Baltimore D. Construction of a retrovirus packaging mutant and its use to produce helper free defective retrovirus. *Cell* 1983; **33**: 153–159

Mulligan RC, Howard BH, Berg P. Synthesis of rabbit beta-globin in cultured monkey kidney cells following infection with a SV40 beta-globin recombinant genome. *Nature* 1979; **277**: 108–114

Partridge TA, Morgan JE, Couton GR, Hoffman EP, Kunkel LM. Conversion of mdx myofibres from dystrophin-negative to -positive by injection of normal myoblasts. *Nature* 1989; **337**: 176–179

Szczylik C, Skorski T, Nicolaides NC, Manzella L, Malaguarnera L, Venturelli D, Gewirtz AM, Calabretta B. Selective inhibition of leukaemia cell proliferation by BCR-ABL antisense oligodeoxynucleotides. *Science* 1991; **253**: 562–565

Tabin CJ, Hoffman JW, Goff SP, Weinberg RA. Adaptation of a retrovirus as a eukaryotic vector transmitting the herpes simplex thymidine kinase gene. *Molec Cell Biol* 1982; **2**: 426–436

Wigler M, Pellice A, Silverstein S, Axel R. Transfer of single-copy eukaryotic genes using total cellular DNA as donor. *Cell* 1978; **14**: 725–731

Chapter 9
Some broader applications

Though most emphasis has been given to the medical applications of recombinant DNA technology, the picture would not be complete without at least a mention of a few other areas in which it is proving to be particularly valuable. Their discussion of perforce must be a little brief and therefore superficial, though hopefully sufficient to whet the reader's appetite.

EVOLUTION AND POPULATION GENETICS

DNA studies permit detailed comparisons to be made of gene structure in different groups of organisms. In this way it is possible to study changes in gene structure which occurred during evolution. This is sometimes referred to as **molecular evolution**. Molecular analysis is also providing a unique way of determining the taxonomic relationships between various groups of animals, and an additional tool for studying evolution and racial diversity in humans.

Molecular evolution

How genes evolve is a fascinating area in its own right and the globin genes provide an ideal system for studying molecular evolution. It seems the α-globin and β-globin gene clusters separated some 500 million years ago. Subsequently, both clusters have evolved by a series of tandem globin gene duplications which then diverged both in sequence and developmental expression to give the current organization. It appears that over millions of years of evolution no intervening sequences were gained or lost by active globin genes, which indicates the incredible stability of this gene complex, perhaps because of its considerable physiological importance. By comparing restriction maps of intergenic DNA it seems that these regions evolved very slowly. This suggests they have been somehow constrained during evolution, from which it follows that they cannot be entirely functionless ('junk').

Interestingly, legumes possess a protein called **leghaemoglobin** which is structurally and functionally similar to animal myoglobin and haemoglobin – so similar in fact that it may well have been derived relatively recently from insect globin sequences, perhaps by an insect-borne plant pathogenic virus.

Taxonomic studies

DNA technology can also provide additional information on the relationships between various genera and species. A useful taxonomic tool, though admittedly somewhat crude and certainly not as precise as DNA sequencing or detailed gene analysis, is **DNA–DNA hybridization** (DNA annealing). Essentially this consists of allowing single-stranded DNA from different species to reassociate. Any mismatching due to evolutionary divergence will reduce the bonding strength between the two DNA molecules and so when the hybrid is heated to separate the two strands, the temperature required to do this will be **lower** than for double-stranded DNA from the same species. A lowering of the dissociation temperature by 1 °C is equivalent to about 1% difference in nucleotide sequences between two DNA molecules. This technique had been used as a measure of 'genetic distance' between taxonomic groups. For example, this is proving a valuable tool in studying the relationships between different species of birds which can be a particularly difficult problem for ornithologists.

Similar studies have revealed that modern man and apes are much more closely related than was once believed. The DNA of man and the higher primates (chimp and gorilla) differ by only 1%. This is a very much smaller difference than would have been expected and is the sort of difference seen in closely related animals such as dogs and foxes, horses and zebras, walruses and sealions. Based on such data, Gribbin and Cherfas have argued, in a very readable way, that the higher primates diverged perhaps no more than four million years ago, and that the common ancestor may have been more man-like than ape-like, and perhaps even bipedal. Interestingly, Dr Mary Leakey had at that time recently discovered an early human ancestor at Laetoli in Tanzania, dating from about $3\frac{1}{2}$ to 4 million years ago, which was clearly bipedal. Along with palaeontologists, anatomists and biochemists, it seems likely that molecular biologists will have much useful information to contribute to the details of human evolution.

DNA polymorphisms

In the past a variety of determinants have been studied in order to estimate, for example, the degree of relatedness between different human populations. These have included various anatomical measurements (physical anthropology) as well as characteristics such as colour blindness and the ability to taste phenylthiocarbamide (PTC). More recently a number of biochemical traits have also been studied from this point of view, including blood groups, plasma proteins (e.g. haptoglobins, transferrins, Gc groups), red cell enzymes (e.g. acid phosphatases, lactate dehydrogenases) and HLA types. Most of these traits constitute what are called 'genetic polymorphisms'. A genetic polymorphism may be defined as the occurrence of two or more alleles at a given locus which are relatively common in the population and therefore serve as useful markers for studying various aspects of population structure, such as inbreeding and migration, etc. In the medical literature, light has been thrown on the structure of certain human populations by also studying the incidence of various genetic disorders (e.g. Eriksson *et al.*, 1980; McKusick, 1978).

A great deal has been written about the results of such studies and the interested reader will find useful summaries in Cavalli-Sforza and Bodmer (1971).

The new technology now offers a way of studying human population genetics at the molecular level from DNA sequence analysis and restriction site polymorphisms. The latter have been found to occur in intergenic regions, intervening sequences, within pseudogenes and within the coding regions of functional genes. Some of these polymorphisms are said to be public because they occur in all racial groups, and the frequency of the least common allele is usually greater than 5%. The remaining polymorphisms are said to be private in that they are only found in a certain population. Again most information on DNA polymorphisms in humans has so far come from studies on globin genes. As discussed earlier, sickle cell anaemia is a serious disease which occurs widely throughout Africa and parts of southern Europe, the Middle East and Asia, and there has been much speculation as to whether the sickle cell mutation had a single or multiple origin. The study of DNA polymorphisms has begun to throw some light on this problem. Kan and Dozy in 1978 found that using a β-globin gene probe and after digesting DNA with the enzyme *Hpa*I, in Caucasians the normal gene was contained in a 7.6 kb fragment. However, in American Negroes the normal β-globin gene also occurred in a variant 7.0 kb fragment and when present the sickle cell β-globin gene usually occurred in a 13.0 kb fragment. These variations are due to a polymorphism at the *Hpa*I site some 5 kb from the 3' end of the β-globin gene (Fig. 9.1).

The linkage with the *Hpa*I site provides a means of diagnosing sickle cell anaemia *in utero* but has also been of significance to human population genetics as described here. The frequency of the 13.0 kb fragment in the American Negro population is shown in Table 9.1.

It should be noted that in this case the linkage is with a specific allele and so close that the term **linkage disequilibrium** is used. Such **allele linkage** is quite common and other examples have been found in the haemoglobinopathies and cystic fibrosis.

Though the sickle cell gene is usually carried on a 13.0 kb fragment, occasionally it may be carried on a 7.6 kb fragment. There are two possible explanations for this situation: either a single sickle cell mutation occurred on a chromosome carrying the 13.0 kb recognition site but that subsequently crossing-over occurred with a chromosome carrying the 7.6 kb recognition site, or separate mutations occurred at different times on different chromosomes bearing either a 13.0 kb or 7.6 kb

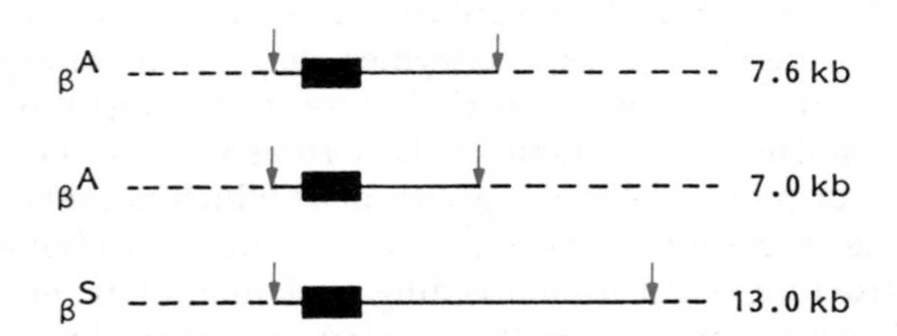

Figure 9.1 DNA fragments containing the β-globin gene generated by the enzyme *Hpa*I.

Table 9.1 *Hpa*I fragment sizes and haemoglobin A and S genes in American Negroes. (Based on data from Kan and Dozy, 1980, with permission)

	Frequency of association	
*Hpa*I fragment size	A gene	S gene
7.6	0.88	0.31
7.0	0.09	0.01
13.0	0.03	0.68

recognition site. Studies of this DNA polymorphism in different populations have revealed that the two *Hpa*I polymorphic sites associated with the sickle cell mutation have different geographical distributions, which suggests that the mutation has arisen as separate events. In fact it has been estimated that up to as many as 10 different sickle cell mutations may have occurred over the past hundred generations. Because of the selective advantage enjoyed by individuals with the sickle cell trait in areas affected by falciparum malaria, once the mutation occurred on a chromosome bearing a closely linked restriction site, the association would tend to be favoured. The phenomenon has been referred to as the **hitch-hiker effect**. An extension of the molecular approach in studying the origin of particular mutations is to consider restriction site polymorphic haplotypes. Here a mutation with a selective advantage would favour the haplotype in which it occurred and studies of β-globin haplotypes also suggest a multiple origin for the sickle cell mutation.

The example of sickle cell anaemia illustrates how studies of DNA polymorphisms may throw light on the origin and subsequent diffusion of a mutation through a population, and the study of restriction site polymorphic haplotypes in different populations may prove particularly helpful in this regard. A dozen or so β-globin haplotypes have now been identified and where a particular haplotype occurs in different populations it carries different β-thalassaemia mutations.

The analysis of the mini-satellites previously described have also been very informative in evolutionary studies, particularly in non-human species. An interesting example relates to the programme of conservation, through captive breeding, of the Puerto Rican parrot. By 1975 the Puerto Rican parrot had been reduced to only about 13 birds and, in order to conserve the species, a few founders were taken from the wild to establish a captive breeding colony. Unfortunately, the number of successful breeding pairs in captivity has been low and the programme has not been as successful as the closely related Hispaniolan parrot. Brock and White carried out a genetic study to examine the level of relatedness of the founder birds, by measuring the levels of band sharing in DNA fingerprints. The Puerto Rican parrots were found to be more closely related than the Hispaniolan birds (unrelated individuals shared on average the same number of bands as second degree relatives of the Hispaniolan birds). The severe inbreeding may well be partly responsible for the low number of breeding pairs and it provides an opportunity for selecting pairs which are more distantly related by choosing couples with the lower levels of band sharing. This provides just one example of a technology adopted enthusiastically by zoologists and population geneticists.

Mitochondrial DNA

So far this discussion of evolution and population genetics has been concerned with nuclear DNA, but mitochondrial DNA has also been extensively studied from this point of view. Mitochondrial DNA is arranged in small circular units located in the cytoplasm with thousands of identical copies per cell. Most of the proteins in mitochondria are coded by nuclear DNA, but a few are exclusively coded by mitochondrial DNA; in humans mitochondrial genes code for the three largest subunits of cytochrome c oxidase as well as cytochrome b, which are involved in important (cellular) oxidative processes.

Mitochondrial DNA has a number of advantages over nuclear DNA as an indicator of evolutionary and population changes. It is readily accessible and small, being about 16.5 kb in length (but very much bigger in higher plants), and has been completely sequenced in man. Because of its abundance it is the easiest to amplify by PCR in archaeological and historical studies. Since its inheritance is exclusively maternal (page 106), in any given population all the mitochondrial DNA will be of the original maternal type and individual female lineages may be traced back over hundreds of generations. Further, it has evolved at least ten times more rapidly than nuclear DNA and extensive polymorphisms in mitochondrial DNA have been found in different human populations. For these various reasons mitochondrial DNA therefore provides an extremely valuable marker for studying human population genetics and a very useful tool for determining evolutionary relationships among closely related species and even among individuals within a species.

From the patterns of restriction sites in mitochondrial DNA it has been possible to construct a phylogeny of human racial groups. An example of how this is achieved is given in Fig. 9.2.

The phylogeny was constructed by relating each pattern through single site changes. The results indicate that the pattern found in 12.5% of Oriental and 4% of Bantu samples might be the ancestral type, which is consistent with other restriction site data, and suggests that the various racial groups may have diverged from an Asian origin. These particular data have been selected because they provide a simple illustration of how information on restriction site patterns in mitochondrial DNA can be used in constructing phylogenies. However, using five different restriction enzymes no less than 35 distinct mitochondrial DNA types have been found! In this case three presumed ancestral types were identified and the resultant phylogeny was far more complicated than the example given in Fig. 9.2, which perhaps would be expected.

Mitochondrial DNA analysis has been used to address the issue of the prehistoric colonization of the South Pacific islands of Oceania (and the spread of genetic disease such as thalassaemia which accompanied it). DNA samples taken from

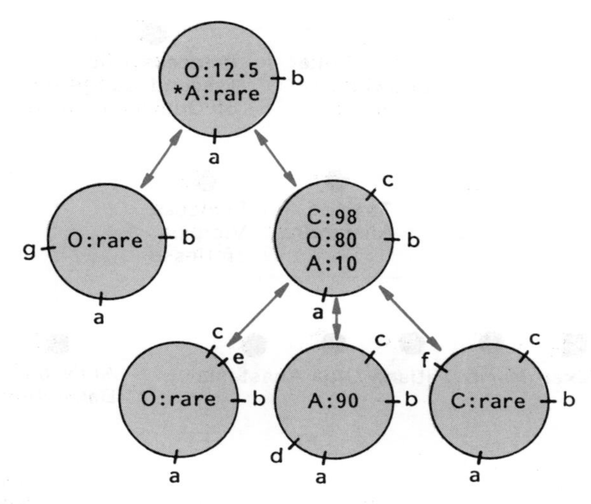

Figure 9.2 Phylogeny based on mitochondrial *Hpa*I restriction site patterns. Numbers refer to the frequencies (%). Restriction sites are indicated as a to g; C = Caucasian; O = Oriental; A = African; *A = Bantu. (From Denaro *et al.*, 1981, with permission.)

skeletons found buried in various sites in Oceania have suggested that settlers arrived from Melanesia as well as South-East Asia.

With the advent of the PCR, the science of archaeobiology has expanded enormously but a single example, which has captured the public imagination, as well as having important international diplomatic ramifications, will be described. From sketchy historical accounts, nine skeletons found in a shallow grave in Ekaterinburg, Russia, were tentatively identified as the remains of the last Tsar of Russia and his family. The group was believed to contain the Tsar, Tsarina, three of their five children, three servants and the Royal Physician, Dr Bodkin. Sex testing was performed by PCR amplification of a short Y specific sequence from extracted DNA. Gold fillings in teeth were thought to indicate an aristocrat rather than a servant. DNA extracts were typed for five simple tandem repeats, each with at least five alleles. The pattern of inheritance of the repeats readily **excluded** all but two of the individuals from being the parents of the children. The fact that all three children had inherited alleles from the likely Tsar and Tsarina in a strictly Mendelian fashion suggested strongly that this was a family group but did not itself prove that the family was the Russian Royal Family. For this part of the study resort was made to the mitochondrial gene variation. Samples were obtained from living relatives of the Tsar and Tsarina who were related through exclusively **maternal** inheritance (Fig. 9.3) including Prince Philip, Duke of Edinburgh. Sharing of polymorphic patterns makes it extremely likely that the skeletons are indeed those of the Tsar and Tsarina, and appropriate reburial is planned in St Petersburg.

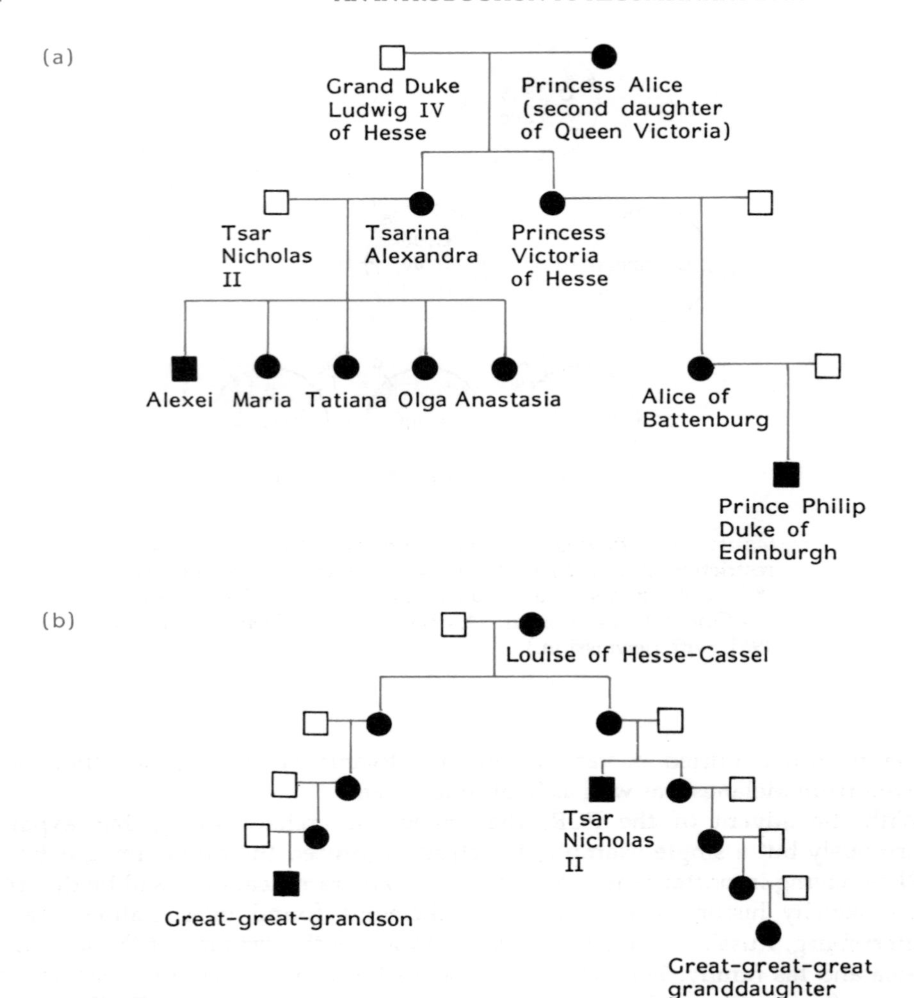

Figure 9.3 (a) Lineage of Tsarina Alexandra, showing relationship to HRH Prince Philip (Duke of Edinburgh). (b) Lineage of Tsar Nicholas II, showing relationship to two relatives tested. Samples from Prince Philip, the great-great-grandson and great-great-great-granddaughter of Louise of Hesse-Cassel were analysed.

AGRICULTURAL APPLICATIONS

Recombinant DNA technology is likely to benefit animal husbandry through the production of vaccines and the application of methods for transferring genes for commercially important traits such as milk yield and butter fat, though this will be difficult because most of these traits are multifactorial. The sort of areas in which the new technology may be expected to benefit arable farming is in developing methods for transferring genes for nitrogen fixation, improving photosynthesis (and therefore yield), resistance to pests, pathogens and herbicides, and tolerance to frost, drought and increased salinity. Again the problem is that most of these characteristics are controlled by many genes which have not yet been identified. Nevertheless, in recent years there have been many exciting developments in plant genetic engineering, quite as dramatic as in other areas. Tobacco plants, for example, have been induced to produce bean protein (phaseolin), anti-viral genes transferred into cotton, and genes for antibiotic resistance have been transferred as well as genes which can be switched on by light and off by darkness, an important step towards regulating the function of genetically engineered plants.

SUMMARY

- In this chapter an attempt has been made to indicate some broader applications of recombinant DNA technology. Molecular studies in various animals as well as in man are providing interesting information on changes in gene structure during evolution (molecular evolution), taxonomic relationships and human population genetics and evolution.
- Studies of mitochondrial DNA are proving particularly useful in relation to racial diversity in man.
- In agriculture perhaps the most important applications are in attempting to transfer genes for nitrogen fixation to cereal crops, though there are formidable problems.

REFERENCES AND FURTHER READING

Textbooks and review articles

Cavalli-Sforza LL, Bodmer WF. *The Genetics of Human Populations*. Freeman, San Francisco, 1971

Emery AEH. *Methodology in Medical Genetics*, 2nd edn. Edinburgh: Churchill Livingstone, 1986

Eriksson AW, *et al.* (eds). *Population Structure and Genetic Disorders*. London: Academic Press, 1980

McKusick VA (ed). *Medical Genetic Studies of the Amish*. Baltimore and London: Johns Hopkins University Press, 1978

Research publications

Brock MK, White BN. Application of DNA fingerprinting to the recovery program of the endangered Puerto Rican parrot. *Proc Natl Acad Sci USA* 1992; **89**: 11121–11125

Denaro M, Blanc H, Johnson MJ, Chen KH, Wilmsen E, *et al.* Ethnic variation in *Hpa*I endonuclease cleavage patterns of human mitochondrial DNA. *Proc Natl Acad Sci USA* 1981; **78**: 5768–5772

Gill P, Ivanov P, Kimpton C, Piercy R, Benson N, Tully G, Evett I, Hagelberg E, Sullivan K. Identification of the remains of the Romanov family by DNA analysis. *Nature Genetics* 1994; **6**: 130–137

Kan YW, Dozy AM. Antenatal diagnosis of sickle-cell anaemia by DNA analysis of amniotic fluid cells. *Lancet* 1978a; ii: 910–912

Kan YW, Dozy AM. Polymorphism of DNA sequence adjacent to human β-globin structural gene; relationship to sickle mutation. *Proc Natl Acad Sci USA* 1978b; 75: 5631–5635

Kan YW, Dozy AM. Evolution of the haemoglobin S and C genes in world populations. *Science* 1980; 209: 388–391

Orkin SH, Antonarakis SE, Kazazian HH. Polymorphism and molecular pathology of the human β-globin gene. *Prog Haematol* 1983; 13: 49–73

Chapter 10
Problems and future of recombinant DNA

The advent of recombinant DNA technology has frequently been described as heralding a new and revolutionary phase in biology, and few would dispute this. No other area of science has received so much public attention with the publication of important findings being announced on radio and TV bulletins concurrent with publication. However, it has also raised a number of problems, including fears of possible biohazards, legal issues concerned with priorities and patents, and quite significant ethical issues. There have been a number of suggestions that the technology has outpaced the moral discussions.

BIOHAZARDS

From the outset considerable concern was expressed about the possible biohazards of this new technology by both scientists and non-scientists. Might organisms emerge which would prove resistant to all known antibiotics or carry cancer genes which might then spread throughout the community? Would the potential benefits be outweighed by these kinds of hazard? On the other hand, others argued that the risks had been greatly exaggerated and did not want limits imposed on research. The controversy was hotly debated on both sides of the Atlantic, and though much of the heat has now dissipated it might be instructive to review what has come to be called the 'Recombinant DNA Debate.

The debate

From the time when the double-helical structure of DNA was first elucidated in 1953, and for nearly twenty years afterwards, a great deal was learned about molecular genetics, largely through the study of *E. coli* and bacteriophages. Much was theoretical and seemed to bear little relevance to important practical problems. Some molecular biologists began to turn their attention increasingly to studying animal tumour viruses in the belief that this could help our understanding of cancer in humans. Fears began to be expressed, however, of the possible biohazards of such research at a time when interest in ecology and pollution was increasing anyway. It was in this climate that the first experiments on recombinant DNA were beginning to be carried out, and the story of the subsequent debate probably begins in the summer of 1971. Janet Mertz, then working as a young 'post-doc' in Paul Berg's laboratory at Stanford University, California, described at a scientific meeting at Cold Spring Harbor an experiment in which it was intended to insert DNA from SV40 into a laboratory strain of *E. coli*. The idea of using such a system to clone 'foreign' DNA in *E. coli* was certainly exciting, but it posed serious problems. *E. coli* flourishes in the human intestine and is ubiquitous, and SV40 was known to cause tumours in certain animals. If, therefore, such genetically engineered organisms escaped from the confines of the laboratory, might they then infect people outside and increase their chances of developing cancer? The experiments were postponed. Fears continued to grow, and in January 1973 a meeting was held in Asilomar in California to discuss the matter, and the battle lines between the two opposing factions became clearly defined. In July 1974 the now famous 'Berg letter' was published in the journal *Science*. It was signed by 11 internationally famous scientists, all renowned for their work in molecular biology. In this letter the signatories explained that through recombinant DNA technology it was possible to create novel types of infectious DNA elements and that the biological properties of such elements could not be predicted with certainty. They therefore recommended that there should be a moratorium on all such experiments until the potential hazards could be better evaluated. The need for caution was echoed the following year in the Ashby Report in the United Kingdom (HMSO Cmnd 5880, 1975), and a second Asilomar Conference was organized. Out of this conference developed the concepts of physical and biological containment. By physical containment is meant limiting the spread of potentially dangerous microorganisms by using specially constructed laboratories and techniques. Biological containment, on the other hand, involves the use of microorganisms which have been attenuated ('crippled') in some way, so that they cannot survive outside the confines of laboratory culture conditions. Bodies were set up both in the United States, through the National Institutes of Health, and in the United Kingdom (Genetic Manipulation Advisory Group – GMAG) to define various levels of containment and provide advice and guidelines for work in this field.

Physical containment can range from Category I (P1 in the United States), where this is minimal and involves the observance of the usual care which has to be exercised in any microbiological laboratory, to Category IV (P4 in the United States), where this is maximal and of an exceptionally high order. To effect biological containment, special microorganisms were developed which possessed various mutations which reduced considerably their likelihood of surviving outside the laboratory.

A number of safer vectors have been developed such as the Charon phages and the plasmid pAT153 which is non-mobilizable. Attempts are being made to quantify the risks associated with various experiments in order to determine the appropriate level of physical containment. For example, the use of the plasmid pAT153 with *E. coli* strain MRC8, where expression of an inserted DNA sequence is not sought but is used to generate a probe for prenatal diagnosis, would require no more than Category I containment or good microbiological practice. But to use a cloning system for the effective expression of a dangerous toxin would require the highest category of containment. Possible impact on the environment is now also seen as an important consideration.

However, in the last few years, most have been won over by the proponents of the new technology. There are several reasons for this. Firstly, as we have seen, most human genes contain introns, the precursor mRNA of which has to be removed and the remaining precursor mRNA from the exons spliced together if a normally functioning protein is to be produced. No mechanisms exist for doing this in bacteria and so any human gene which is introduced into such an organism will only be transcribed and expressed if it is specifically 'tailored' by the experimenter. Secondly, the risks associated with recombinant DNA experiments have proved to be considerably lower than was at first imagined. Thirdly, all laboratories in which this work is carried out have to be registered and all research workers are periodically examined. Despite the hundreds of laboratories and thousands of scientists involved in this work, there have been no reports of anything untoward happening. Nevertheless, care needs to be exercised continually, especially when dealing with potential pathogens and their toxins, and in industry where production is on a vastly different scale from the research laboratory.

LEGAL PROBLEMS

Now that the fears of recombinant DNA work have been largely dispelled, in their wake other issues have come into prominence. In particular, the new technology has generated legal problems concerned with priorities and patents for various aspects of the technology and the creation of new life forms. In the past scientists have usually been content to report their findings without giving any thought to commercial considerations. However, no doubt because of the immense financial gains to be made in the various industrial applications of the technology, it is perhaps understandable that research organizations, and a few individuals, have sought to recoup some of their financial investments in research by applying for patents.

The first patent issued in this field was granted by the United States Supreme Court in June 1980 to Dr Ananda Chakrabarty for a specially created *Pseudomonas* bacterium which would break down oil slicks. The argument was accepted that this discovery was not nature's handiwork (which would not be patentable) but his own. Cohen and Boyer also applied for a patent to cover the basic process involved in generating recombinant DNA molecules and this was also granted on behalf of Stanford University in December 1980.

The polymerase chain reaction is also protected by a patent and licence fees may be paid for use of the process and the *Taq* polymerase. More controversially, the teams at the Universities of Michigan and Toronto who discovered the cystic fibrosis gene are requesting a licence fee from laboratories carrying out screening and testing. The intention is that the charities who paid for so much of the initial research should benefit so that they can invest in further research, for example into treatment. However, the perception of this differs somewhat on either side of the Atlantic depending on whether the health care system is based on private insurance or a National Health Service. Perhaps the most disruptive example of patenting was the attempt, firstly by scientists at the National Institutes of Health in the USA and rapidly afterwards by the UK Medical Research Council, to patent short pieces of sequence from random cDNAs. This was deemed unusual and inappropriate in that no function for the gene was shown, nor the full sequence produced and indeed there was no real proof that the sequence came from a gene. Such patenting has, therefore, been abandoned.

ETHICAL ISSUES

To some extent the new technology presents ethical problems which have already been debated. In the case of prenatal diagnosis the same ethical problems of abortion remain, but the ability to carry out chorion biopsy in the first trimester of pregnancy may make the procedure more acceptable in future. The use of the technology to treat patients on an experimental basis before adequate animal studies have been completed is not a new problem for medical ethicists. DNA technology does, however, raise several new issues with important ethical implications. The possibility of treating an early human embryo by intranuclear injection of mouse DNA seems unlikely to be seriously countenanced in the near future. Quite apart from the risks to the viability of the embryo, even if it survived and the technique of inserting a specific DNA sequence was successful, the idea of changing the genetic constitution of an individual would be repugnant to many. In any event this is still a very long way from being a reality for reasons discussed earlier.

Genetic screening also poses many questions. Population screening for cystic fibrosis would appear to be desirable and achievable. Technically, there is the possibility of screening large numbers of people by DNA samples obtained from mouth washes with the possibility of detecting 85% of carriers and 70% of carrier couples (see page 130). However, many questions arise, not least the time at which the screening should be carried out. If children were tested at birth it would have the advantage that virtually everyone would be included. However, it raises the issue of whether it is an individual's own right, rather than their parents', to decide whether they wish to be tested. In general, it is considered that minors under the age of 18 should not be tested until, and if, they decide so for themselves. Children could be tested while at school, and pre-conception screening would seem desirable, but again the question of legal consent arises, together with the problem that 15 or 16 year olds will have different levels of sexual experience and they may be stigmatized amongst their group if the results become publicly known. A further possibility is testing in early pregnancy. It can be argued that only **couples** proceeding with a pregnancy need to be tested, but a major disadvantage presents itself. If both partners test positive for a known cystic fibrosis mutation, their risk of having an affected child is 1 in 4. However, as already explained, a person who does not show one of the common mutations could still be a carrier of a rare mutation. If neither partner tests positive, the chance of them having an affected child is very low indeed, but if one tests positive and one negative then the chance is 1 in 620. As the chance without testing is 1:2300, this is actually a **raised** risk, and they will require counselling by a trained professional. As 1:24 couples will come in this category, a large number of people will have had unnecessary anxiety at a sensitive time and large amounts of counselling time will be used to discuss a risk which is, compared to other risks in pregnancy, small. The first pilot studies to test the various models of screening delivery are being conducted.

Again the scenario will vary somewhat depending on the legal situation. In more litigious countries medical professionals may be wary of giving too reassuring information, fearing that they could be sued if an affected child is born to a couple who felt they had been told there was little or no risk. On the other side of the coin, couples may try to sue if they find they have not been offered a test which could have avoided the birth of an affected child. Other objections have been put forward. It has been suggested that children born with cystic fibrosis, when prenatal testing was available, may be treated as 'guilty' in some way. It has even been suggested that research into treatment may decline if the birth of affected individuals is seen as largely avoidable. This last possibility has been completely discounted by the wealth of physiological, biochemical and therapeutic studies which have been undertaken in the years since the gene was discovered and have given us confidence that more effective treatments will soon be available.

The ethical and practical issues are complex enough for cystic fibrosis which is serious and has simple genetics. In other disorders, for example late onset diseases such as motor neurone disease or Huntington's chorea, the issues become far more contentious. Should abortion be allowed for an illness which will not develop until after many years of normal life (e.g. breast cancer, Huntington's) or may never develop at all (e.g. schizophrenia or coronary heart disease)? If someone has been tested, who is entitled to the information? Should an insurance company be allowed to request the information? Should an individual be allowed to tell an insurance company that they are at **reduced** risk? Will an underclass develop who are unable to get jobs or insurance? These issues are being actively discussed now, but in some countries, particularly in Germany, they are regarded as particularly sensitive because of previous eugenics abuse. The European Union, which has made a major funding commitment to the genome mapping project, has insisted that a proportion of grants go to ethical and legal studies. Hopefully, a generation of lawyers will arrive who have the scientific training to understand the technology and the issues.

Recombinant DNA technology challenges the Copernican view of our world by giving man the ability to alter its natural structure and order. As Ledley has argued, in debating the implications of recombinant DNA technology, these are not resolved by simply pronouncing judgements on its efficacy and safety, but rather we should 'begin to explore how society intends to approach the larger issues of man's relationship with the natural structure and order of the universe'.

THE FUTURE

One of the main aims of this book has been to emphasize the general principles underlying recombinant DNA technology, and it would seem unlikely that these will continue to change very fast. What is perhaps of interest is to speculate on possible new avenues of knowledge which may be opened up in the future.

Perhaps the most fundamental, and also formidable, problems yet to be solved are concerned with how specific genes are turned on or off in particular tissues and at particular stages of development. How are genes made to function in a coordinated fashion? How, in molecular terms, does the interaction between different cells produce specific cell patterns and cell lineages and eventually the development of tissues and organs? It seems likely that answers to such questions will come from knowing more about the nature and action of tissue-specific enhancers, promoters and other flanking sequences. Not only would such knowledge be of considerable academic interest, it would also help us to understand how hamartomas arise (local malformations due to defective tissue combinations) and how congenital abnormalities are caused by teratogenic agents such as X-rays, chemicals, drugs and certain infections. It would provide a molecular explanation for the phenomenon of apoptosis, or programmed cell death, which occurs in certain tissues and organs during normal development. The involutional changes which take place during the natural ageing process might also be better understood when more is known of how levels of gene activity are controlled.

Answers to some of the problems of cell development are emerging from studying lower organisms such as the fruit fly *Drosophila* and nematode worms, but it will be a long time yet before the findings can be extended to higher forms including man.

Another area in which our current knowledge is woefully deficient is in understanding brain mechanisms and behaviour. As Iversen (1984) has pointed out, at present 'it is clear that ignorance greatly outweighs knowledge in the field of brain chemistry', but until more is known of the molecular basis of normal behaviour, is it likely that disorders such as schizophrenia, depression and drug dependency will ever be completely understood? Perhaps the application of the new technology in some of the ways discussed earlier may prove helpful. However, as in the case of cell development and differentiation, the difficulties of exploring such fundamental problems cannot be underestimated and no single technology alone will be able to provide all the answers. Nevertheless, many fundamental questions in biology may well find solutions only through the new technology.

SUMMARY

- Recombinant DNA technology has doubtless heralded a new and revolutionary phase in biology. However, it has also raised a number of problems. Though the possible biohazards of the technology may have been exaggerated at the beginning, caution is still necessary, especially when dealing with potential pathogens and their toxins, and in industry where production is on a vastly different scale from the research laboratory.
- Legal problems have arisen over priorities and patents, a unique situation being the possibility of patenting new life forms.
- New ethical problems raised by DNA technology include the possibility of perhaps one day altering the germ line through treating human embryos.
- The deliberate contamination of the environment, e.g. by spraying engineered microorganisms onto crops, raises both legal and ethical issues.

- The future applications of the technology, apart from the obvious extensions of existing techniques and approaches, will be in discovering how gene action is controlled in specific tissues and at particular stages of development, and in understanding the molecular basis of human behaviour. These are formidable problems but future developments in these areas will no doubt prove to have extremely important applications in medicine. At a meeting held in Boston to commemorate thirty years' work since the discovery of the double helix, James Watson said: 'If you are young there is really no option but to be a molecular biologist'. For those practising medicine, this is certainly so, at least in thought if not in deed.

REFERENCES AND FURTHER READING

Berg P, Baltimore D, Cohen SN, David RW, Hogness DS, Nathans D, Roblin R, Watson JD, Weissman S, Zinder ND. Potential biohazards of recombinant DNA molecules. *Science* 1974; **185**: 303

Iversen LL. Chemical communications in the brain and mechanisms of action of psychotropic drugs. In: Duncan R, Westwood-Smith M, eds. *The Encyclopaedia of Medical Ignorance*. Oxford: Pergamon Press, 1984, pp. 55–62

Richards JD, Bobrow M. Ethical issues in clinical genetics: A report of a joint working party. *J Roy Coll Physicians London* 1991; 25: 284–288

Watson JD, Tooze J. *The DNA Story – A Documentary History of Gene Cloning*. San Francisco: Freeman, 1981

Williams R. Universal community screening for cystic fibrosis. *Nature Genetics* 1993; 3: 195–201

Glossary

A: abbreviation for adenine.
Adenine: a purine base in DNA and RNA.
Allele (=allelomorph): alternative forms of a gene found at the same locus on homologous chromosomes.
Alu: dispersed, repetitive DNA sequences, about 300 bp long.
Amber codon: (UAG) one of the three stop codons.
Amino acid: an organic compound containing both carboxyl (–COOH) and amino groups ($-NH_2$).
Amniocentesis: procedure for obtaining amniotic fluid and its contained cells for prenatal diagnosis.
Aneuploid: a chromosome number which is not an exact multiple of the haploid number, i.e. $2N-1$ or $2N+1$, where N is the haploid number of chromosomes.
Annealing: occasional alternative term for hybridization.
Antibody (=immunoglobulin): a serum protein which is formed in response to an antigenic stimulus and reacts specifically with this antigen.
Anticipation: increasing severity of disease with passing generations.
Anticodon: triplet in tRNA complementary to a particular codon in mRNA.
Antigen: a substance which elicits the synthesis of antibody with which it also reacts specifically.
Antiparallel: opposite orientation of the two strands of a DNA duplex, one running in the 3' to 5' direction, the other running in the 5' to 3' direction.
Ascertainment: the finding and selection of families with an hereditary disorder.
Association: the occurrence of a particular allele in a group of patients more often than can be accounted for by chance.
Assortive mating (=non-random mating): the preferential selection of a spouse with a particular phenotype.
Autoradiography: detection of radioactively labelled molecules on X-ray film.
Autosome: any chromosome other than the sex chromosomes. In man there are 22 pairs of autosomes.
B DNA: natural right-handed form of DNA.
B lymphocytes: lymphocytes which secrete antibodies.
Bacteriophage (=phage): a virus which infects bacteria.
Base: short for nitrogenous base in nucleic acid molecules (A = adenine; T = thymine; U = uracil; G = guanine; C = cytosine).
Base pair (bp): a pair of complementary bases in DNA (A with T, G with C).
Blunt-ended ligation: a reaction which joins two DNA molecules at their blunt ends.
Blunt termini (blunt ends): DNA molecule with double-stranded (blunt) ends.
C: abbreviation for cytosine.

Cap: a methylated structure added to the 5' end of mRNA.
Capsid: protein coat of a virus particle.
Carcinogen: an agent which causes cancer.
CAT box: a conserved, non-coding, so-called 'promoter' sequence about 80 bp upstream from the start of transcription.
cDNA: single-stranded DNA complementary to an mRNA.
CentiMorgan (cM): unit used to measure map distances. Equivalent to 1% recombination (crossing-over).
Centromere: the point at which the two chromatids of a chromosome are joined, and the region of the chromosome which becomes attached to the spindle during cell division.
Chiasma: the cross-configuration of the chromatids of homologous chromosomes during the first meiotic division. Represents the point at which crossing-over occurs.
Chorion: layer of cells covering the fertilized ovum.
Chromatid: during cell division each chromosome divides longitudinally into two strands or chromatids which are held together by the centromere.
Chromosomal aberration: an abnormality of chromosome number or structure.
Chromosomes: thread-like, deep-staining bodies situated within the nucleus and composed of DNA and protein.
Chromosome walking: sequential isolation of clones which carry overlapping DNA sequences so as to traverse part of a chromosome.
Cistron: the smallest unit of genetic material which is responsible for the synthesis of a specific polypeptide.
Clone: all the cells derived from a single cell by repeated mitoses and all having the same genetic constitution (*see* **DNA cloning**).
cM: abbreviation for centiMorgan.
Coding strand: DNA strand with a base sequence comparable to mRNA.
Codominance: when both alleles are expressed in the heterozygote.
Codon: a sequence of three adjacent nucleotides (triplet) which codes for one amino acid or chain termination.
Cohesive termini (cohesive ends): DNA molecule with single-stranded ends with exposed (cohesive) complementary bases.
Congenital: any abnormality, whether genetic or not, which is present at birth.
Consanguineous marriage: a marriage between 'blood relatives', i.e. between persons who have one or more common ancestors, usually a marriage between first cousins.
Controlling elements: mobile genetic elements (transposons) in maize.
Cosmid: plasmid DNA packaged *in vitro* into a phage.
Cos site (=cohesive end site): cohesive ends of a phage DNA molecule.
Cross-over (=recombination): the exchange of genetic material between a pair of homologous chromosomes
Cytoplasm: the ground substance of the cell in which are situated the nucleus, endoplasmic reticulum and mitochondria, etc.
Cytoplasmic inheritance: inheritance of genes located in mitochondria or chloroplasts.
Cytosine: a pyrimidine base in DNA and RNA.

Deletion: loss of a DNA sequence (or part of a chromosome).
Denaturation: change of a protein to an inactive form. Also disruption of DNA duplex into two separate strands (=melting).
Derepression: activation or induction of gene transcription.
DGGE (denaturing gradient gell electrophoresis): method of detecting changes in DNA sequence by passing through a polyacrylamide gel with an increasing gradient of denaturant.
Diploid: the condition in which the cell contains two sets of chromosomes. Normal state of somatic cells in man, where the diploid number (2*N*) is 46.
DNA (=deoxyribonucleic acid): the nucleic acid in which genetic information is stored (apart from some viruses).
DNA cloning: production of many identical copies of a defined DNA fragment.
DNA polymerase: an enzyme which catalyses the synthesis of double-stranded DNA from single-stranded DNA.
DNase: an enzyme which produces single-stranded nicks in DNA and is used in nick translation.
Dominant: a trait which is expressed in individuals who are heterozygous for a particular gene.
Downstream: refers to sequences located farther along in the direction of transcription.
Duplex: double-stranded nucleic acid molecule.
Endonucleases: enzymes which cleave bonds within nucleic acid molecules.
Endoplasmic reticulum: a system of minute tubules within the cytoplasm.
Enhancer: DNA sequence which increases transcription of a related gene.
Enzyme: a protein which acts as a catalyst in biological systems.
Eukaryotes: higher organisms with a well-defined nucleus.
Exon: region of DNA which generates that part of precursor RNA which is **not** excised during transcription and forms mRNA and thus specifies the primary structure of the gene product.
Exonucleases: enzymes which catalyse the removal of nucleotides from the ends of a DNA molecule.
First-degree relatives: closest relatives; i.e. parents, offspring and sibs.
Five prime (5') end: the end of a DNA or RNA strand with a free 5' phosphate group. The end of a gene at which transcription begins.
Frameshift mutations: mutations which change the reading frame in which triplets are translated into protein.
G: abbreviation for guanine.
Gamete: a germ cell (sperm or ovum) containing a haploid (*N*) number of chromosomes.
Gene: a part of the DNA molecule which directs the synthesis of a specific polypeptide chain. It is composed of many codons. When the gene is considered as a unit of function in this way, the term cistron is often used.
Genome: all the genes carried by a cell.
Genomic DNA: DNA sequences in the chromosome.
Genotype: the genetic constitution of an individual.
Guanine: a purine base in DNA and RNA.

Haploid: the condition in which the cell contains one set of chromosomes. Normal state of gametes in man where the haploid number (*N*) is 23.

Haplotype: a group of closely linked alleles which are inherited together as a unit.

Hemizygous: a term used when describing the genotype of a male with regard to an X-linked trait, since males have only one set of X-linked genes.

Heritability: the proportion of the total variation of a character attributable to genetic as opposed to environmental factors.

Heteroduplex: hybrid duplex formed by pairing between two different DNA molecules which are not completely complementary.

Heterogametic sex: the sex which produces gametes of two types. In humans, the male is the heterogametic sex because he produces X- and Y-bearing sperms.

Heterozygote (=carrier): an individual who possesses two different alleles at one particular locus on a pair of homologous chromosomes.

Histone: type of protein rich in lysine and arginine found in association with DNA.

Hogness box (=TATA box): a conserved, non-coding, so-called 'promoter' sequence about 30 bp upstream from the start of transcription.

Homogametic sex: the sex which produces gametes of only one type. In humans the female is the homogametic sex because she produces only X-bearing ova.

Homograft: graft between individuals of the same species but with different genotypes.

Homologous chromosomes: chromosomes which pair during meiosis and contain identical loci.

Homopolymer tailing: addition of a number of bases of the same type to the 3' ends of a DNA molecule using the enzyme terminal transferase.

Homozygote: an individual who possesses two identical alleles at one particular locus on a pair of homologous chromosomes.

'Housekeeping' genes: genes expressed in all cells because they provide basic functions required by all cells.

Hybrid: the progeny of a cross between two genetically different organisms.

Hybrid-arrested translation (HART): technique used to identify DNA sequences by hybridization with RNA and studying the products of *in vitro* protein synthesis.

Hybridization: the pairing of RNA and DNA strands or two different DNA strands.

Hybridoma: cell line, produced by fusion of myeloma cells and B lymphocytes, which will grow in culture indefinitely and produce monoclonal antibodies.

Immunoglobulin: *see* **Antibody**

Imprinting: parental influences on gene expression.

Incompatibility: a donor and host are incompatible if the latter rejects a graft from the former.

Inducer: small molecule which interacts with a regulator protein and triggers gene transcription.

Induction: switching on of transcription.

Insertion: additional DNA sequence within the genome.

***In situ* hybridization**: hybridization with an appropriate probe carried out directly on a chromosome preparation or histological section.

Intervening sequence: *see* **Intron**.

Intron (=intervening sequence): region of DNA which generates that part of precursor RNA which is excised during transcription and does not form mRNA and therefore does not specify the primary structure of the gene product.

IS (=insertion sequence): a small bacterial transposon.

Isochromosome: a type of chromosomal aberration in which one of the arms of a particular chromosome is duplicated because the centromere divides transversely and not longitudinally during cell division. The two arms of an isochromosome are therefore of equal length and contain the same genes.

Karyotype: the number, size and shape of the chromosomes of a somatic cell. A photomicrograph of an individual's chromosomes arranged in a standard manner.

Kb: abbreviation for kilobase.

Kilobase: 1000 base pairs (bp).

Library: set of cloned DNA fragments which together represent the entire genome or the transcription of a particular tissue.

Ligase: enzyme used to join DNA molecules.

Ligation: formation of phosphodiester bonds to link two nucleic acid molecules.

Linkage: two genes at different loci on the same pair of homologous chromosomes are said to be linked.

Linkage disequilibrium: the association of two linked alleles more frequently than would be expected by chance.

Linker molecule: synthetic oligonucleotide containing a site for a restriction enzyme.

Liquid (solution) hybridization: hybridization between nucleic acid molecules performed in solution.

Locus: the site of a gene on a chromosome.

Lod score: logarithm of the odds score – a measure of the likelihood of two loci being within a measurable distance of each other.

LTR (=long terminal repeat): a sequence directly repeated at both ends of a retroviral DNA.

Lysis: destruction of bacteria by infective phage.

Lysogeny: survival of a phage in a bacterium as an integrated prophage in the bacterial genome.

M13: a single-stranded DNA phage.

Major histocompatibility complex: a set of genes on the short arm of chromosome 6 which determine the immune response.

Maternal inheritance (=cytoplasmic inheritance): gene complex on chromosome 6, which includes the HLA genes and various complement components (factor B, C2 and C4) and is concerned with transplantation rejection.

Map distance: measure of the distance between linked gene loci expressed in centiMorgans (cM).

Marker: DNA fragment of known size used to calibrate an electrophoretic gel. Also a loose term for a DNA polymorphism or any allele of interest in an experiment.

Maternal inheritance (=cytoplasmic inheritance): transmission of a trait exclusively through maternal relatives.

Meiosis: the type of cell division which occurs during gametogenesis and results in halving of the somatic number of chromosomes so that each gamete is haploid.

Melting: denaturation of DNA.

Metaphase: the stage of cell division when the chromosomes line up on the equatorial plate and the nuclear membrane disappears.
MHC: *see* **Major histocompatibility complex**
Missense mutation: a mutation which results in a change of an amino-acid-specifying codon.
Mitochondria: small circular DNA units within the cytoplasm.
Mitosis: the type of cell division which occurs in somatic cells.
Monoclonal antibody: pure antibody produced by a single clone of cells.
Mosaicism: the existence of tissues with different chromosome complements within the same individual.
Multifactorial: inheritance controlled by many genes with small additive effects (polygenic) plus the effects of environment.
Multiple alleles: the existence of more than two alleles at a particular locus in a population.
Mutagen: an agent which causes mutations.
Mutation: a change in the genetic material, either of a single gene (point mutation) or in the number or structure of the chromosomes. A mutation which occurs in the gametes is inherited; a mutation which occurs in the somatic cells (somatic mutation) is not inherited.
Mutation rate: the number of mutations at any one particular locus which occurs per gamete per generation.
Myeloma: tumour cell line derived from a lymphocyte.
N-terminal: the end of a polypeptide that has an amino acid with a free amino group.
Nick: a break in the sugar–phosphate backbone of a DNA or RNA strand.
Nick translation: *in vitro* method used to introduce radioactively labelled nucleotides into DNA.
Non-disjunction: the failure of two members of a chromosome pair to disjoin (=separate) during cell division so that both pass to the same daughter cell.
Nonsense mutation: a mutation producing any one of three codons (UAG, UAA or UGA) which cause termination of protein synthesis.
Nonsense suppressor: mutant tRNA which suppresses a nonsense mutation.
Northern blot: technique for transferring RNA fragments from an agarose gel to a nitrocellulose filter on which they can be hybridized to a complementary DNA.
Nucleolus: a structure within the nucleus concerned with protein synthesis.
Nucleoside: the base–sugar moiety of a nucleic acid molecule.
Nucleosome: DNA–histone subunit of a chromosome.
Nucleotide: the base–sugar–phosphate moiety of a nucleic acid molecule.
Nucleus: a structure within the cell which contains the chromosomes and nucleolus.
Ochre codon: (UAA), one of the three stop codons.
Oligonucleotide: a chain of, literally, a few nucleotides.
Oligoprobe: abbreviation for oligonucleotide probe.
Oncogene: cancer gene.
Oncogenic: literally cancer causing.
Operator gene: a gene which switches on adjacent structural gene(s).
Operon: complete unit of bacterial gene expression consisting of a regulator gene(s), control elements (promoter and operator) and adjacent structural gene(s).

pBR322: a standard plasmic cloning vector.
Penetrance: the proportion of heterozygotes who express a trait even if mildly.
Peptide bond: chemical bond between the carboxyl bond (–COOH) group of one amino acid and the amino ($-NH_2$) group of another.
Phage: abbreviation for bacteriophage.
Phage lambda (**λ**): a phage cloning vector.
Phenotype: the appearance (physical, biochemical and physiological) of an individual which results from the interaction of environment and his genotype.
Plaque: clear area in a plated bacterial culture due to lysis by phage.
Plasmid: small, circular DNA duplex capable of autonomous replication within a bacterium.
Pleiotropy: a gene with multiple effects is said to be pleiotropic.
Point mutation: single base pair change.
Polyadenylation: addition of a sequence of polyadenylic acid (poly A-tail) to the 3' end of mRNA after transcription.
Poly A tail: *see* **Polyadenylation.**
Polygenic: literally many genes.
Polymerase chain reaction: amplification of a specific sequence by repeated rounds of oligonucleotide binding and extension.
Polymorphism: put simply, the occurrence of two or more alleles at a locus which are relatively common in the population. (There are other complicated and more embracing definitions.)
Polypeptide: an organic compound consisting of three or more amino acids.
Polyploid: any multiple of the haploid number of chromosomes (e.g. 3*N*, 4*N*, etc.).
Polysome (=polyribosome): a group of ribosomes associated with the same molecule of messenger RNA involved in protein synthesis.
Post-translational modification: various modifications of proteins which occur after their synthesis.
Primer: a short sequence of bases which provides a start for the synthesis of a deoxyribonucleotide chain.
Proband (=index case): an affected individual (irrespective of sex) through whom the family came to the attention of the investigator. Propositus if a male; proposita if a female.
Probe: a labelled DNA fragment which will hybridize with, and thereby detect and locate, complementary sequences among DNA fragments on, say, a nitrocellulose filter. Labelled RNA probes are also available.
Processing: alterations of RNA which occur during transcription including splicing, capping and polyadenylation.
Prokaryotes: lower organisms with no well-defined nucleus.
Promoter: recognition sequence for binding of RNA polymerase.
Prophage: phage DNA that has been integrated into the host DNA.
Protein: a complex organic compound composed of hundreds or thousands of amino acids.
Proto-oncogene: a normal gene with the potential to become activated and hence a cancer gene (oncogene).
Provirus: retroviral DNA that has been integrated into the host DNA.
pSC101: a plasmid cloning vector.

Pseudogene: DNA sequence homologous with a known gene but is functionless.
Purine: a nitrogenous base with fused five- and six-member rings (adenine, guanine).
Pyrimidine: a nitrogenous base with a six-member ring (cytosine, uracil, thymine).
R loop: configuration produced by the displacement of a single-stranded DNA loop from a DNA duplex over a region of hybridization with complementary RNA.
Recessive: a trait which is expressed in individuals who are homozygous for a particular gene but not in those who are heterozygous for this gene.
Recombinant: a vector containing a 'foreign' DNA sequence. Also used for an individual who represents a cross-over (recombination) between two linked loci.
Recombination: cross-over between two linked loci.
Recombination fraction (θ): frequency of recombination.
Regulator gene: a gene which synthesizes a repressor substance which inhibits the action of a specific operator gene.
Repetitive DNA: DNA sequences of variable length which are repeated up to 100 000 (middle repetitive) or over 100 000 (highly repetitive) copies per genome.
Repression: inhibition of gene transcription.
Restriction endonucleases: group of enzymes which cleave DNA. Type II enzymes cleave DNA at sequence specific sites.
Restriction fragment: DNA fragment produced by a restriction endonuclease.
Restriction fragment length polymorphism (RLFP): polymorphism due to the presence or absence of a particular restriction site.
Restriction map: linear arrangement of various restriction enzyme sites.
Restriction site: base sequence recognized by a restriction endonuclease.
Retrovirus: RNA virus which replicates via conversion into a DNA duplex.
Reverse transcriptase: an enzyme which catalyses the synthesis of DNA from RNA.
RFLP: *see* **Restriction fragment length polymorphism**.
Ribosomes: minute spherical structures in the cytoplasm. They are rich in RNA and the seat of protein synthesis.
RNA (=ribonucleic acid): The nucleic acid which is found mainly in the nucleolus and ribosomes. **Messenger** RNA transfers genetic information from the nucleus to the ribosomes in the cytoplasm and also acts as a template for the synthesis of polypeptides. **Transfer** RNA transfers activated amino acids from the cytoplasm to messenger RNA.
RNA polymerase: an enzyme which catalyses the synthesis of RNA in transcription.
S1 nuclease: an enzyme which degrades single-stranded DNA.
Segregation: the separation of alleles during meiosis so that each gamete contains only one member of each pair of alleles.
Selection: ability of certain genotypes in the population to survive and reproduce. Also used for methods for enhancing the survival of required cell colonies in tissue culture.
Sex chromosomes: the chromosomes responsible for sex (XX in women, XY in men).
Sex linkage: genes carried on the sex chromosomes. Since there are very few Mendelizing genes on the Y chromosome the term is often used synonymously for X-linkage.

Shotgun experiment: cloning of an entire genome in the form of randomly generated DNA fragments.

Sib (=sibling): brother or sister.

Somatic cells: all body cells apart from the germ line.

Southern blot: technique for transferring DNA fragments from an agarose gel to a nitrocellulose filter on which they can be hybridized to a complementary DNA.

Splicing: the removal of introns and joining of exons in RNA during transcription (introns are spliced out; exons are spliced together). Also used for methods which join one DNA molecule to another.

Split gene: a gene containing one or more introns.

SSCP (single-strand conformation polymorphism): different mobilities in polyacrylamide gels, of single-stranded DNA molecules. The conformation depends on the sequence of the DNA segment.

Sticky ends: complementary single strands of DNA.

Stop codon: one of three codons (UAG, UAA and UGA) which cause termination of protein synthesis.

Supercoiling: excessive coiling of a DNA duplex.

Suppression: the elimination of the effects of a DNA mutation without reversing it.

SV40: Simian (monkey) virus 40.

T: abbreviation for thymine.

T lymphocytes: lymphocytes concerned with cellular immunity.

TATA (Hogness) box: a conserved, non-coding, so-called 'promoter' sequence about 30 bp upstream from the start of transcription.

Template: a pattern.

Teratogen: an agent which causes congenital abnormalities.

Terminal transferase: enzyme which catalyses the addition of nucleotides to the 3' ends of DNA.

Termination codon: *see* **Stop codon**.

Three-prime (3') end: The end of a DNA or RNA strand with a free 3' hydoxyl group. The end of a gene at which transcription ends.

Thymine: a pyrimidine base in DNA.

Transcription: the process whereby genetic information is transmitted from the DNA in the chromosomes to messenger RNA.

Transfection: acquisition of new genetic material in eukaryotes by the incorporation of added DNA.

Transformation: in prokaryotes, the acquisition of new genetic material by the incorporation of added DNA. In eukaryotes the conversion of normal cells to malignant cells in culture.

Transgenic: an organism containing genetic material artificially inserted from another species.

Transition: a mutation in which a purine is substituted by another purine, or a pyrimidine by another pyrimidine.

Translation: the process whereby genetic information from messenger RNA is translated into protein synthesis.

Translocation: the transfer of genetic material from one chromosome to another. If there is an **exchange** of genetic material between two chromosomes then this is referred to as a **reciprocal translocation**. If a small fragment is formed (**centric fusion**) which is usually lost, this is then referred to as a **Robertsonian translocation**.

Transposon: mobile genetic element able to replicate and insert a copy at a new location in the genome.

Transversion: a mutation in which a a purine is replaced by a pyrimidine, and vice versa.

Triplet: a series of three bases in a DNA or RNA molecule.

Triplet repeats: runs of many copies of a trinucleotide in a tandem repeat.

Trisomy:a chromosome additional to the normal complement (i.e. $2N+1$) so that in each somatic nucleus one particular chromosome is represented three times rather than twice.

tRNA: small RNA molecules involved in translation.

Tumour suppressor: gene whose normal function is to hold cell division in check.

U: abbreviation for uracil.

Unifactorial (=Mendelizing): inheritance controlled by a single gene pair.

Upstream: refers to sequences located in the opposite direction to transcription.

Uracil: a pyrimidine base in RNA.

Vector: a plasmid, phage or cosmid into which foreign DNA may be inserted for cloning.

Virion: virus particle.

X-linkage: genes carried on the X chromosome are said to be X-linked.

Z DNA: left-handed form of DNA.

Zygote: the fertilized ovum.

Index

Index compiled by Liza Weinkove